Veterinary PCR Diagnostics

Edited By

Chengming Wang

Yangzhou University
Jiangsu, China

Bernhard Kaltenboeck

Auburn University
AL, USA

Mark D. Freeman

Ross University School of Veterinary Medicine
West Indies

CONTENTS

FOREWORD

It was little more than a decade ago that PCR users were given the possibility to follow the amplification cycles in real time. The new machines were able to measure the accumulation of amplified product *via* a fluorescent signal generated by exonuclease digestion of a specifically annealed dual-labeled fluorogenic probe. This meant that users could take a closer look at the kinetics of their amplification reaction, so that, as a consequence, optimization was no longer a matter of gut feeling, but became tangible thanks to the quantitative data provided by the real-time thermocycler. The omission of post-amplification sample handling, such as gel electrophoresis, was another advantage because it significantly reduced the risk of contamination by product carry-over and enabled more rapid and high-throughput testing. Moreover, while conventional PCR would only confirm the presence or absence of a given pathogen in a sample, the real-time technology enabled the investigator to quantitate the amount of this agent.

The quantitation option has opened up new horizons to researchers by facilitating a rigorously quantitative approach to characterize cellular processes, the course of an infection, as well as the spread and dissemination of a pathogen. A whole area of research, *i.e.* transcriptomics, could not have reached the high level it has today without the use of quantitative real-time PCR as a gold standard.

The present book is filling a notable gap. As the amount of scientific publications is steadily increasing, so is the number of PCR-based methodologies and diagnostic assays. Harmonization and standardization of protocols, both in a national and worldwide context, remain important issues on the agenda of diagnostic laboratories. In this situation, there is a necessity to systematically digest the existing literature and give an overview to already specialized users. Furthermore, newcomers need guidance and technical advice before they are able to become productive in the field. This book will be a valuable orientation guide to the community of investigators involved in research and diagnosis of microbial infections, as well as cancer and other diseases that are amenable to nucleic acid-based analysis.

Konrad Sachse
Friedrich-Loeffler-Institut
(Federal Research Institute for Animal Health)
Institute of Molecular Pathogenesis
Germany

PREFACE

A speedy and accurate identification of pathogens is of vital importance for the effective control and management of veterinary infectious diseases. Infectious agents have been traditionally identified with the use of various phenotypic procedures, such as morphological, biochemical and serological assays. However, the phenotypic diagnoses are usually slow and lack proper specificity and sensitivity. The revolutionary invention of the nucleic acid amplification technologies such as PCR allows the detection of pathogens at nucleic acid level, and has played an increasing role in the laboratory diagnosis of infectious diseases. PCR-based technologies offer the ability to detect a single copy of nucleic acid template with supreme sensitivity, specificity, speed and precision for the detection of pathogens. Furthermore, the recent advances in probe chemistries, availability of multiple fluorescent channels in the PCR machines as well as instrumentation automation have facilitated the development of quantitative PCR that provides a convenient platform for high through-put quantitation and differentiation of pathogens in clinical specimens of veterinary medicine.

We are very fortunate and honored to have international specialists as chapter contributors in their respective specialty of veterinary PCR diagnostics. Mrs. Salma Sarfaraz at Bentham Science Publishers is a constant source of encouragement and discipline for the production of this book. This book provides a reliable, convenient and comprehensive reference on molecular detection and identification of pathogens of veterinary significance. Chapter **1** (Loftis *et al.*) outlines the principles of real-time PCR, and chapter **2** (Marijke) describes a practical guiding standard that can be used in different steps of the design and validation of in-house developed real-time PCR assays. Chapter **3** (Lilenbaum *et al.*) focuses on molecular diagnosis of veterinary bacterial infections, exemplified with *Brucella* sp., *Leptospira* sp., *Mycobacterium bovis*, *Staphylococcus aureus* and *Mycoplasma* sp. Chapter **4** (Shaheen *et al.*) provides valuable knowledge on the molecular mechanisms of drug resistance in microbial pathogens and the potential advantages and disadvantages of PCR-based methods. Chapter **5** (Li *et al.*) highlights the diagnosis of viral disease in livestock and companion animals with PCR, and Chapter **6** (Wu *et al.*) focuses on the application of real-time PCR for detection and differentiation of significant parasites in veterinary medicine and public health, exemplified in protozoa, helminthes and arthropods. In chapter **7**, Gentilini *et al.* focus their attention on PCR applications for the diagnosis and prognosis of cancer in pets which are already currently available, albeit not diffusely, at both academic and private laboratories around the world.

Chengming Wang
Yangzhou University
Jiangsu, China

Bernhard Kaltenboeck
Auburn University
AL, USA

Mark D. Freeman
Ross University School of Veterinary Medicine
West Indies

List of Contributors

Sudhir K. Ahluwalia
Banfield Pet Hospital, 10 Traders Way
Salem, MA 01970, USA

Dawn M. Boothe
Department of Anatomy, Physiology & Pharmacology
Auburn University, Auburn
AL, USA

Claudia Calzolari
Department of Veterinary Medical Sciences
University of Bologna
Italy

Mark D. Freeman
Ross University School of Veterinary Medicine
Basseterre
St. Kitts, West Indies

Fabio Gentilini
Department of Veterinary Medical Sciences
University of Bologna
Italy

Kirsten Jaegersen
Ross University School of Veterinary Medicine
Basseterre
St. Kitts, West Indies

Bernhard Kaltenboeck
Department of Pathobiology
Auburn University
AL, USA

Yihang Li
Geriatrics Research, Education and Clinical Center
Department of Veterans Affairs Palo Alto Health Care System
California, USA

Walter Lilenbaum
Biomedical Institute
Fluminense Federal University
RJ, Brazil

Amanda D. Loftis
Ross University School of Veterinary Medicine
Basseterre
St. Kitts, West Indies

Raymaekers Marijke
Clinical Laboratory, Jessa Hospital
Site Virga Jesse, Belgium

Rajesh Nayak
Division of Microbiology
National Center for Toxicological Research
US Food and Drug Administration, Jefferson
AR, USA

Renata F. Rabello
Biomedical Institute
Fluminense Federal University
RJ, Brazil

Will K. Reeves
USAF School of Aerospace Medicine
Wright-Patterson Air Force Base
Ohio, USA

Boakai K. Robertson
Department of Biological Sciences
Alabama State University, Montgomery
AL, USA

Konrad Sachse
Friedrich-Loeffler-Institut
(Federal Research Institute for Animal Health)
Institute of Molecular Pathogenesis
Jena, Germany

Bashar W. Shaheen
Division of Microbiology
National Center for Toxicological Research
US Food and Drug Administration, Jefferson
AR, USA

Rubens C. da Silva Dias
Biomedical Institute
Federal University of State of Rio de Janeiro
Rio de Janeiro
RJ, Brazil

Maria E. Turba
Department of Veterinary Medical Sciences
University of Bologna
Italy

Robert Villafane
Department of Biological Sciences
Alabama State University, Montgomery
AL, USA

Chengming Wang
School of Veterinary Medicine
Yangzhou University
Jiangsu, China

Hongzhuan Wu
Department of Biological Sciences
Alabama State University, Montgomery
AL, USA

Veterinary PCR Diagnostics

CHAPTER 1

Principles of Real-Time PCR

Amanda D. Loftis[1*] and Will K. Reeves[2]

[1]*Ross University School of Veterinary Medicine, Basseterre, St. Kitts, West Indies and* [2]*USAF School of Aerospace Medicine, Wright-Patterson Air Force Base, Ohio, USA*

Abstract: Compared with traditional PCR assays, diagnostic assays based upon real-time PCR technology have increased speed and dynamic range; in addition, they enable quantitative analysis of gene copies and have the potential for increased specificity when nucleic acid probes are used. Optimized real-time PCR assays can also be highly sensitive, detecting as few as 1-10 copies of a target gene in a nucleic acid sample. Adopting real-time PCR in a diagnostic laboratory requires an understanding of these assays, including both the benefits and drawbacks unique to this technology. An overview of real time PCR applications is presented here, with an emphasis on practical issues that might affect implementation of real-time PCR testing in a diagnostic laboratory. Increased cleanliness and process controls are required in the laboratory, to prevent contamination of sensitive real-time PCR. Nucleic acid extraction procedures, using one of the many available chemistries, should be carefully optimized for reproducible, efficient extraction of nucleic acids that are free of PCR inhibitors. Reverse transcription of RNA adds an additional variable that can affect quantitative data. For the assay itself, different options have been developed for the detection of products in real-time, including dye-based assays, hydrolysis probes, and hybridization probes. Different options and the benefits and drawbacks of each are discussed. Finally, specific applications for real-time quantitative PCR assays in diagnostic laboratories are highlighted.

Keywords: Real-time PCR; polymerase chain reaction; clinical laboratory techniques; reverse transcription; DNA; RNA; nucleic acid probes.

1. INTRODUCTION TO REAL-TIME PCR

The polymerase chain reaction (PCR) is a biochemical process that copies and amplifies DNA using a thermally stable DNA polymerase. Real-time PCR is increasingly being adopted by diagnostic laboratories, both in the human and veterinary medical fields. Since the first description of real-time PCR in 1992 [1], the field has expanded rapidly, with significant improvements in chemistry, analysis of data, and availability and affordability of real-time PCR platforms. Several reviews and book chapters are available which discuss the application of real-time PCR in various fields; however, the authors often assume that the reader already has a background in the field and focus on recent improvements in the technology or on specialized issues such as template normalization, the optimal equations to calculate reaction efficiency, *etc.* While these issues are important for the advanced user, fewer works are available that are intended for a laboratory that is considering adding real-time PCR to their services. This chapter is intended to introduce the field, to provide an overview of the different applications and chemistries available for diagnostic real-time PCR, and to discuss the pitfalls and benefits of real-time PCR from a practical standpoint. For the context of this chapter, the terms "diagnosis" or "diagnose" will be used to refer to the detection of a pathogen.

Real-time PCR was initially developed as a variation on conventional PCR assays. In conventional PCR, a DNA template is mixed with DNA polymerase, nucleotides, and other components of the PCR; thermocycling is performed for a total of 30-45 cycles; and the resultant product is then tested for the presence of a distinct DNA amplicon. The DNA template can be derived from pure cultures, clinical or diagnostic specimens, intact organisms, or environmental samples.

RNA has several biological and biochemical differences from DNA that make it more difficult to handle in

---

*Address correspondence to Amanda D. Loftis: Ross University School of Veterinary Medicine, P.O. Box 334, Basseterre, St. Kitts, St. Kitts and Nevis; Tel: +1-869-465-4161 x408; E-mail: adloftis@gmail.com, aloftis@rossvet.edu.kn

Chengming Wang, Bernhard Kaltenboeck and Mark D. Freeman (Eds)

the field or laboratory, and the polymerase enzymes used in PCR do not amplify RNA. Testing for RNA is only possible following a reverse transcription step to convert the RNA to ssDNA. The resulting ssDNA is referred to as complementary DNA or cDNA. Many infectious diseases in animals and humans are caused by viruses, a significant proportion of which have a RNA genome. In addition, the detection of messenger RNA (mRNA) is required for gene expression studies.

The primary difference between real-time and conventional PCR assays is that products of a real-time reaction are measured in "real time", as the PCR reaction is being performed, rather than after the reaction is complete. In real-time PCR, a fluorescent detector is added to the PCR, and the fluorescence of each sample is measured during each cycle of amplification. The use of fluorescence to detect PCR amplicons improves the dynamic range for real-time PCR. In the earliest form of real-time PCR, ethidium bromide, a fluorescent dye which intercalates into dsDNA, was used for detection [1], several additional dsDNA dyes, with improved sensitivity and reduced toxicity, were subsequently validated and have replaced the use of ethidium bromide (*e.g.*, [2, 3]). In dye-based real-time PCR systems, melt curve analyses are used to confirm the identity of PCR amplicons by measuring the melting temperature (T_m) of the resultant product; after the amplification is complete, samples are cooled and then slowly heated, while fluorescence is monitored to determine the temperature at which the dsDNA-binding dye is released (*e.g.*, [3, 4]).

As the field advanced, probe-based methods, which couple the hybridization of a fluorescently labeled oligonucleotide probe to the amplification of the real-time PCR product, were developed [5]. Probes are generally labeled with a fluorescent reporter on one end and a fluorescence quencher on the other; during the process of amplifying the target DNA, degradation or hybridization of the probe physically separates the reporter from the quencher and generates a fluorescent signal. Probe-based methods are more specific, analogous to combining a conventional PCR with Southern blotting. The use of probes allows the additional option of combining (multiplexing) two or more different assays, using probes with different fluorescent reporters, in a single reaction tube. Alternately, one set of PCR primers can be used, with multiple different probes, to detect and discriminate between genotypes or pathogens (*e.g.*, [6]). For increased stringency, a pair of probes can be used, both labeled with fluorescent compounds, in which the transfer of fluorescent energy between the probes when in close physical proximity generates a signal (LightCycler® or HybProbe systems). In systems in which probes hybridize but are not hydrolyzed, melt curve analysis can be used to determine the temperature at which the probes dissociate from the template.

Whereas conventional PCR is typically qualitative in nature, yielding only positive and negative results, real-time PCR adds the potential for quantitative analysis. As amplification progresses over several cycles, the fluorescence generated by the dye or probe increases until this fluorescence rises significantly above baseline. This is recorded as the threshold cycle or C_T; the greater the starting quantity of target DNA, the lower the C_T. This principle is the basis for quantitative real-time PCR analysis. In its simplest form, several dilutions of a quantified DNA template are used to generate a standard curve, after $\log_{10}$ transforming the number of gene copies, from which the number of gene copies in an unknown sample are extrapolated. Various methods of normalizing these quantitative data are used, for different applications, based upon the quantity of starting DNA or host DNA, the amplification of a housekeeping gene, *etc.* The former methods are used to reduce variations in data caused by differences in the amount of template used in the reaction. Housekeeping genes are typically used in RNA quantification assays, to control for variability in the efficiency of the reverse transcription step.

From a practical standpoint, real-time PCR eliminates the necessity of post-PCR processing, reducing labor, minimizing the time required to obtain a result, and preventing contamination of the laboratory from amplicon handling. However, the sensitivity of real-time PCR makes these assays more sensitive to PCR inhibitors that may not be removed by the nucleic acid extraction step, requiring careful validation of extraction procedures as well as the real-time PCR assay(s). These assays are also sensitive to false positive results caused by contamination of water sources, primers, or probes, or by aerosolization of highly concentrated templates. As a result, real-time PCR should be prepared in a designated area, separate from that used for nucleic acid extraction, which is regularly cleaned with chemicals or irradiated with UV light. Cross-contamination during the process of preparing the reactions is further minimized by maintaining a "clean to dirty" work flow: prepare the master mix first, then handle the unknown DNA specimens, and, finally, add positive controls, starting with the least concentrated template and finishing with the most concentrated. There are also several chemical options available to prevent amplicon cross-contamination [7]. Most accredited diagnostic laboratories have standard operating procedures to reduce the risk of contamination.

Nomenclature for real-time PCR can be confusing. Various authors use the abbreviation "RT-PCR" to refer to reverse transcriptase PCR, for conventional PCR detection of RNA, or to refer to real-time PCR assays for the detection of DNA. The term "qPCR" is usually used to refer to quantitative real-time PCR assays for DNA detection, whereas "qRT-PCR" or "RT-qPCR" may be used to refer to quantitative real-time, reverse transcriptase assays for RNA. Some authors use "qPCR" to refer to any real-time PCR assay, whether quantitative or not. The abbreviation "RRT-PCR" is also used to refer to real-time, reverse transcriptase PCR, which may or may not be quantitative in nature. To minimize confusion, none of these abbreviations will be used in this chapter.

2. PREPARATION OF TEMPLATE

The primary concerns for nucleic acid template quality for real-time PCR are similar to those for other applications: yield, purity, and lack of inhibitors. The primary difference is in the increased emphasis on consistency of yield between samples and in the removal of all PCR inhibitors: quantitative studies require that the efficiency of extraction is similar for all samples and controls included in the study, and the improved sensitivity of real-time PCR for detection of template is accompanied by an increased sensitivity to PCR inhibitors. Inhibitors of PCR can co-purify with the nucleic acids, may fail to be removed by insufficient washing, or can be introduced from the laboratory environment; some common inhibitors of real-time PCR include: heme, heparin, EDTA, ethanol, and compounds found in feces [3, 8, 9]. Similarly, nucleic acid templates and oligonucleotides for real-time PCR should be prepared in Tris-HCl buffer or water, but Tris/EDTA (TE) should be avoided due to the potential inhibition by EDTA.

Prior to the actual use of real-time PCR, time should be spent validating the nucleic acid extraction method with the sample type of choice. Yield will be negatively affected by exposure of the nucleic acids to nucleases (*e.g.*, DNases or RNases), either during sample collection, processing, or storage. As a precaution, all equipment and containers that come in contact with purified nucleic acids should be nuclease free. Nucleic acids will bind to silicates, including glass, so glassware is often avoided. If glass beads are used to disrupt a sample, the amount of time the beads are left in contact with these materials should be minimized.

Overall nucleic acid yield of the extraction method should be assessed quantitatively, with a DNA-binding dye (for example, PicoGreen) or UV spectrophotometry; UV absorbance is less accurate than DNA binding dyes, due to the overlap in absorbance spectra between DNA, RNA, and protein. Efficiency of the nucleic acid extraction and the presence of inhibitors can be simultaneously assessed by adding a known quantity of purified template DNA, typically a quantified positive control standard, to a control sample. The extraction and real-time PCR testing are completed, as would be done for an unknown sample, and the number of gene copies that are detected is compared to the quantity used to prepare the original sample. To assess the presence of inhibitors, without the added variability of extraction efficiency, add the purified template DNA directly to prepared extracts and to a control containing only buffer and/or carrier nucleic acids; compare the quantity of the control that is detected in the two samples.

2.1. Methods for Preparation of DNA Templates

In general, the effects of DNase activity during sample collection can be minimized by keeping cellular membranes intact. DNases are used by cells, in a highly regulated and restricted fashion, to degrade DNA as part of normal cellular metabolism. Eukaryotic cells may have DNase in nuclear granules or lysosomes, and bacteria use methylation and other strategies to protect their cytoplasmic genome from restriction enzymes and other cytoplasmic nucleases. Depending upon the type of sample and duration of storage, samples may be refrigerated (short-term), frozen, stored in >70% ethanol or isopropanol, or dried completely on filter paper. Freeze-thaw, mechanical disruption, or chemical disruption of cellular membranes will permit the endogenous DNases to contact genomic DNA. These methods of disruption are commonly included in protocols to release DNA into solution; however, in some instances (*e.g.*, freeze-thaw), the release of DNase is an unintended consequence of sample storage or processing. When cell membranes are disrupted for any reason, chemical or physical methods should be used to prevent DNases from degrading the template. Examples include: using proteases, such as Proteinase K, to degrade nucleases; slowing enzyme activity by working with samples at 4°C; or including chaotropic salts, such as guanidinium thiocyanate, in extraction buffers.

Once samples have been properly collected and stored, several methods are available for preparation of DNA templates; selection of the appropriate method may depend upon budget, time constraints, and the number, type, and volume of samples. The most commonly used methods for DNA preparation are based upon one of four systems: biphasic purification, silica-gel based column purification, magnetic bead purification, and boiling with chelation of PCR inhibitors. Different methods may be better suited for different applications; published papers are available that compare DNA extraction methods for common sample types (*e.g.*, [10-17]). A brief overview of each type of chemistry follows.

Biphasic purification, using phenol/chloroform, phenol/isoamyl alcohol/chloroform, or proprietary kit-based systems, can be used to prepare high-quality templates for real-time PCR, as long as the pellets are thoroughly dried following alcohol precipitation and the volume of reagents is suitable for the sample being tested. This approach has been used successfully on a variety of sample types, including semi-solid and viscous materials, and offers flexibility in the scale of extractions. In most cases, biphasic purification protocols are generated in-house and should be carefully validated prior to use.

Several kits are available that use silica-gel based columns for purification of DNA samples, either in a single tube or 96-well plate format; these kits are available for different types of samples (cell free, blood, tissue, bacteria, feces, *etc.*) and are provided with protocols specifically designed for the type of sample. In most cases, as long as the sample is of appropriate type and size for the kit, washing is complete, and all ethanol is removed, by centrifugation or vacuum, prior to elution of the template, DNA produced by these methods is suitable for real-time PCR. These kits are primarily suitable for use on liquid (or liquefied) samples that will not clog the silica gel matrix.

Magnetic bead purification systems are designed for higher-throughput laboratories and offer a method to produce purified DNA in an automated or semi-automated fashion (*e.g.*, [18]). These systems are designed for high-throughput work, such as that performed in diagnostic laboratories, but smaller models are also available. The yield of DNA is often lower than that seen with other systems, but magnetic beads offer an increased capacity to remove PCR inhibitors by extensive washing. Commercial applications allow several options including RNA, DNA, or total nucleic acid extraction kits. As with other commercial kits, the proper kit type must be matched with the type of sample.

Finally, samples can be boiled to lyse cells and release DNA and then subjected to chelation using a resin such as Chelex-100; this method does not purify the DNA, *per se*, but can produce a suitable template that is free of PCR inhibitors. These methods have been applied extensively to forensic samples, bacterial colonies, and other specimens of small volume [19-22]. Additionally, chelation can be applied to DNA samples that have already been purified using another method, to remove inhibitors. Heparin co-purifies with DNA, and chelation is the only method that has been proven to remove this inhibitor from DNA samples [23].

2.2. Preparation of Templates from RNA

Because RNA is rapidly degraded by free ribonucleases (RNases) in biological samples, proper collection and preservation of samples is the first step in successful reverse transcription PCR. Cells use RNA to manage most biochemical reactions and cellular regulatory activities; as a result, they must be able to rapidly degrade RNA, and all cells contain RNases [24]. Even the touch of a human fingertip, or breathing into a sample, can deliver enough RNase to degrade RNA. RNA can be preserved by physical and chemical methods. RNA is stable in most biological samples at ultralow temperatures, and storage in liquid nitrogen, on dry ice, or in a -80°C freezer should suffice for most preservation. Additionally, some viruses are stable, even infectious, when stored in a dry sample. A wide range of chemicals can be used to preserve RNA. These include fixatives such as Carnoy's solution, which can be made in the laboratory, and proprietary RNA stabilizing solutions (*e.g.*, [25]). Many of these preservatives will degrade DNA, so they should not be used on samples where detection of DNA is also required. The method of choice should be related to the properties of the agent or the suspected agent and the possible need to isolate viable virus from preserved tissues.

Since most diagnostic reverse transcriptase PCR is aimed at identifying viruses, the initial step in sample collection should focus on the diagnostic need. Freezing will preserve RNA but could damage and inactivate some types of viruses and cellular pathogens. Many chemical preservatives will completely inactivate a virus but preserve the RNA. If a diagnostic laboratory needs to culture virus from a sample, this must be considered prior to specimen collection. For example, West Nile Virus can be detected by reverse transcriptase PCR in dead mosquitoes stored dry at room temperature for weeks [26], however, the virus is inactivated and cannot be cultured. Bluetongue virus, on the other hand, remains stable in blood in a refrigerator for months to years but is degraded by freezing [27]. In a best case situation, the field collection of a sample suspected to contain an RNA virus should be tailored to the probable viral agents, or else multiple preservation methods can be used.

Viral agents with RNA genomes are different from cellular pathogens. The biochemical structure of the virion, the type of nucleic acid in the genome, and the structure of those nucleic acids, varies between viral families. DNA viruses have a DNA-based genome and can be treated just like cellular pathogens for DNA extraction. RNA viruses can have single or double stranded RNA, the genome can be continuous or segmented, and the single stranded RNA genomes can be in the sense or nonsense orientation. Some basic background knowledge of the viral agent being detected is important. Basic surveillance for a known agent is achieved through an easier process than the identification of a true unknown. Identification of the virus family in the sample by electron microscopy or serology can be helpful in narrowing down the options.

RNA extraction methods are similar to those used for DNA, with biphasic extractions, silica-gel based systems, and magnetic bead platforms available for use with RNA. In addition to manual extraction methods, many commercial kits are available for preparation of RNA. The primary difference when working with RNA is the increased susceptibility of RNA to either enzymatic or biochemical degradation. Only RNase-free plasticware, glassware, and reagents should be used, and laboratory staff need to observe strict cleanliness, including wearing gloves to prevent skin RNase enzymes from contaminating specimens. Chemicals used throughout RNA extraction and for storage of purified RNA often include inhibitors of RNase enzyme activity. Similar to DNA extraction, RNA extraction methods have been reviewed and compared in the published literature (*e.g.*, [28, 29]). Advances in RNA extraction and purification are rapidly ongoing, so focusing on a single kit or technique will be outdated quickly.

One critical issue with RNA extraction and purification is the sample preservation method. Frozen or dry samples can be processed with little to no special treatment. Samples stored in preservatives, such as Carnoy's or highly saline commercial preservatives, will need to be washed or cleaned prior to RNA extraction, because these fixatives can inhibit or inactivate the extraction process. Additionally, one major difference between most RNA extraction techniques and the extraction of RNA from a virus is the size of the RNA fragments. RNA molecules used by cells, such as tRNA, mRNA, and rRNA, are relatively short compared to many viral genomes. Viral RNA can contain thousands of base pairs with complicated tertiary structure. Depending on the downstream use of the RNA, some extraction methods will provide higher quality products. To extract viral RNA, the method must be applicable to larger RNA molecules. If the purpose of the assay is to quantify mRNA transcripts, the method should be optimized for shorter RNA molecules. Always read the performance standards and manuals of a commercial kit prior to purchase.

Once the extraction process is underway, RNA molecules are susceptible to degradation by ribonucleases. Ribonucleases are released from the host cells and can degrade the products. Commercially available RNA extraction kits often contain RNase inhibitors, ranging from 2-mercaptoethanol to proprietary compounds.

There is no specific need to use an RNase inhibitor, but most samples will be degraded at some point if these inhibitors are not incorporated in the extraction process. 2-Mercaptoethanol, which is used in some widely available kits, is extremely noxious, and care must be taken when working with this chemical or related thiols. A second potential problem is contamination with DNA from the host cells, because of the biochemical similarities between RNA and DNA. DNA contamination can be particularly important when a host cell gene is used as an internal control or when performing quantitative studies of mRNA transcripts.

For applications in which both RNA and DNA copies of a gene exist in the sample and accurate quantitation of the RNA is desired, an extra step must be performed to remove DNA from the extract. Many commercial extraction kits include a DNase as an optional reagent. DNase enzymes may be added after the RNA extraction is complete; the sample is then incubated, followed by heating to inactivate the DNase. Alternately, amplification of eukaryotic DNA might be prevented by specifically designing the PCR primers to prevent amplification of genomic DNA (by intron spanning, *etc.*).

RNA extracts can be stored in a refrigerator or freezer after purification. If any RNases are introduced to the purified products, the RNA can rapidly degrade. Extreme care should be taken to prevent contamination of purified extracts. In addition, repeated freezing and thawing will mechanically shear RNA molecules; multiple freeze-thaw cycles should be avoided. One option for storage and shipment of RNA for viral diagnosis is to immediately reverse transcribe some of the extract using random hexamers and store the cDNA.

2.3. Reverse Transcription of RNA

PCR is a biochemical process for copying DNA. RNA must be converted to DNA prior to detection, using the process of reverse transcription. This biochemical reaction uses an RNA-dependant DNA polymerase to make a single stranded copy of DNA that is complementary to the RNA molecule [30]. The reverse transcriptase enzyme does require a primer; primers that are commonly used include oligo(dT) primers, random primers, or primers specific to the target of choice. Oligo(dT) primers take advantage of the poly-A tail on the 3' end of eukaryotic mRNA and are suitable for assaying mRNA when the primers are located near the 3' end of the gene. Random primers are commercially prepared cocktails of several short primers (6-11 bases) that can anneal to any part of the RNA transcript. Some of these cocktails are truly random, whereas others include primers that are designed to preferentially amplify RNA from specific taxa (enterobacteriaceae, *etc.*). Finally, specific primers can be used to amplify only the target of interest.

A limitation in many commercial reverse transcription kits is the optimal RNA fragment length. RNA purified from single stranded RNA viruses can be reverse transcribed using standard protocols for mRNA in most commercially available kits. Most reverse transcriptases function at a lower temperature than PCR and do not require special denaturation of single stranded RNA. Some RNA viruses have double stranded RNA, which self anneals, and these viruses are generally difficult to reverse transcribe without denaturation. Double stranded RNA can be chemically denatured, but RNA is heat stable and rapid heating to 95°C, followed by cooling on ice or ice-cold ethanol, will suffice for denaturation [31]. For example, double stranded RNA from an Orbivirus (such as Bluetongue Virus or EHDV) must be denatured prior to reverse transcription, but ssRNA from vesicular stomatitis viruses do not.

The RNA genomes of the various families of viruses are poorly conserved between groups [32]. As a result, there are no universal reverse transcription real time PCR assays for viruses. While almost all cellular life shares numerous highly conserved genes in their genomes, such as rRNA genes, there are no conserved genes between the dozens of viral families. When diagnosing a true unknown virus, real-time PCR is probably not a good initial option.

Reverse transcription PCR can be used to detect non-viral pathogens, but the primers and probe must be specifically designed to amplify cDNA from the RNA of the target organism. Some genes in bacteria are not transcribed, and mRNA in eukaryotic cells can have introns that are spliced out. Reverse transcription PCR has some benefits in detecting non-viral pathogens. The number of RNA transcripts probably outnumbers the DNA gene copies found in an active cell. The presence of mRNA in a sample also gives greater evidence that the pathogen is alive, instead of a dead or inactivated environmental contaminant.

Depending upon the application, reverse transcription can be performed on an aliquot of the RNA, and then the resulting ssDNA may be used as the template for several different real-time PCR assays (two-step systems), or the transcription and real-time PCR can be performed sequentially in a single reaction tube (one-step). When reverse transcription is performed in advance and multiple targets are desired, oligo(dT) or random primers are typically used for the reverse transcription reaction. Using one pool of transcribed RNA for all the real-time PCR can minimize quantitative variance caused by run-to-run variation in efficiency of the reverse transcription reaction. The other approach is to perform the reverse transcription and real-time PCR in a single tube, in which one of the PCR primers additionally functions as the specific primer for reverse transcription. In the latter case, both the reverse transcriptase and a "hot start" DNA polymerase are included in the tube; reverse transcription is performed first, typically at temperatures below 60°C, and then heating to 95°C simultaneously degrades the reverse transcriptase and activates the DNA polymerase, with PCR thermocycling following subsequently. This approach is most commonly used when assaying for a limited number of multiplexed targets, for presence/absence experiments, or when a housekeeping gene is included in each reaction tube to normalize the quantitative data. It has the advantage of minimizing handling and reducing the possibility for sample contamination or cross-contamination. Care must be taken in choosing the housekeeping gene to make sure this assay is compatible with that for the pathogen.

Quality control for both the reverse transcription and PCR are critical, and the process is often more complicated when starting with RNA instead of a DNA template. Partial or total failure of either the reverse transcription or PCR can jeopardize the overall performance of the assay. Multiple controls can be incorporated to test the efficacy of both reactions. A well-characterized and highly expressed host regulatory gene is often used as a positive control, which should be detected in any sample from an animal if both the reverse transcription and PCR worked. Likewise, there should be no RNA detected from the negative control. Additional controls can be applied. For example, a traditional PCR product, or a real-time PCR product that uses primers placed on two introns spanning an exon, can detect DNA contamination. A control to measure the efficiency of the PCR step can also be incorporated, consisting of premade cDNA or a synthetic oligonucleotide that corresponds to the primers and probe used in an assay. If this real-time PCR control yields positive results but the reverse transcriptase real time PCR control does not, there is evidence for failure of the reverse transcriptase step. The presence of DNA from the host will make the controls for reverse transcriptase invalid because host DNA will yield positive products.

3. CHOICES FOR DETECTION CHEMISTRY

Detection chemistry is based either upon dyes that bind to double-stranded DNA, including PCR products, or oligonucleotide probes that specifically bind to the target region between the primer annealing sites. In general, dye-based assays are easier to design and optimize, since they have fewer components. These assays are most suitable for broad-range assays that detect a genus or a group of organisms and are unsuitable for multiplexed use. In contrast, probe-based assays can be multiplexed, in which each different probe is labeled with a unique fluorescent reporter; design and optimization are more complex, but the resulting assays can be highly specific, often to the species or strain level. A brief introduction to some of the more common detection chemistries follows; for more detail, several papers have been published which discuss the comparison between, and selection of, detectors for real-time PCR (*e.g.*, [3, 33-37]).

The design of primers and probes for the PCR detection of cDNA templates is different from that for traditional DNA applications. DNA extracts from cells are typically double stranded, and primers can be designed to amplify DNA from either strand. With the exception of dsRNA viruses, the cDNA produced during the reverse transcription from RNA viruses or mRNA will be ssDNA representing only one strand. Careful consideration is needed to avoid making primers or probes that do not anneal in the proper orientation for successful real time PCR. This is particularly true when molecular markers, such as fluorescent molecules or quenchers, must be cleaved from probes during polymerization.

3.1. Detection Based Upon Fluorescent DNA-Binding Dyes

Although early applications used ethidium bromide, dye-based detection is presently based upon more sensitive dyes, especially SYBR Green I, which intercalate into dsDNA and emit fluorescence (Fig. **1A**).

The improved signal : noise ratio of these newer dyes, combined with whole-reaction imaging, using the high-quality digital camera incorporated into the real-time PCR platform, contribute to the improved sensitivity of real-time PCR relative to gel-based detection of PCR products. SYBR Green I is widely used and is available in several ready-to-use, commercially available, kits from various manufacturers. Recent studies show that some newer dyes, while slightly more expensive, may offer better absolute sensitivity and melt curve discrimination; SYTO13 and SYTO82 are two such dyes [2]. Additionally, several proprietary dyes are offered by commercial manufacturers (*e.g.*, BioRad, Promega, Biotium, *etc.*). In most cases, these newer and proprietary dyes are designed to have similar emissions spectra as more traditional dyes, allowing the use of existing filter sets on the real-time PCR platform for detection. For example, SYBR Green I, SYTO13, and FAM (a fluorochrome used in probe-based assays) are detected at the same wavelengths, making it possible to use these dyes in machines with the same detection parameters. With dye-based detection systems, the concentration of dye is constant between platforms and assays, requiring no extra validation of the detection component.

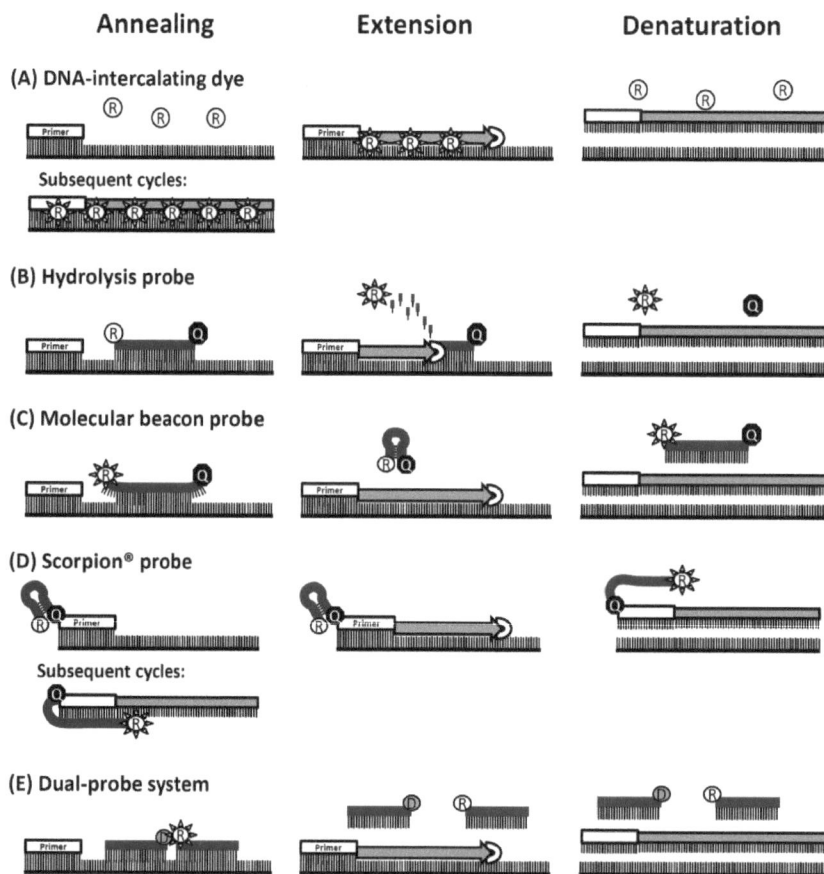

Figure 1: Diagram illustrating five types of detection systems commonly used for real-time PCR assays and their activity during thermocycling conditions used for annealing, extension, and denaturation. For simplicity, only one strand of the template is represented. The reporter fluorophore (R), quencher (Q), and donor fluorophore (D) are included, as appropriate, and distinctions are made between an inactivated reporter and a reporter that is fluorescing. During denaturation, note the lack of specific dye-based fluorescence and the presence of non-specific fluorescence exhibited by molecular beacon and Scorpion® probes; data collection for real-time PCR is typically performed during the annealing stage of the reaction.

When using dye-based detection, careful primer design is essential, as the dye will detect primer dimers and other nonspecific amplicons as well as specific target amplification, creating false positive signals. As an extra precaution, many investigators perform melt curve analysis after the PCR amplification is complete, to establish the temperature at which the dsDNA amplicons dissociate from each other and thus from the DNA-binding dye. Primer dimers should dissociate at significantly lower temperatures than specific amplicons and can be readily discriminated from target amplification. The melting temperature (T_m) should be consistent for each specific amplicon, and in some assays the T_m can be used to discriminate between specific amplicons from related taxa. This provides a putative identification of the target and reduces the possibility of false-positive results. However, optimal primer design should still minimize the formation of primer dimers, since extensive dimer formation decreases the reaction efficiency and, therefore, its sensitivity for target detection.

3.2. Detection Based Upon Labeled Oligonucleotide Probes

In probe-based detection systems, the fluorescent reporter is covalently bonded to an oligonucleotide probe that is designed to anneal to the template between the primers. All of these systems depend upon the transfer of fluorescent energy between two different molecules, either a reporter and a quencher or a reporter and a donor fluorophore. In single-probe systems, the probe is also labeled with a fluorescence quencher; when the reporter and quencher are in close physical proximity, the light omitted by the reporter is absorbed by the quencher and dissipated as heat energy. Fluorescence is recovered when the reporter and quencher are separated. In two-probe systems, each probe is labeled with a different fluorophore, one with a "donor" molecule that emits energy in the excitation spectrum for the "acceptor" fluorophore; the acceptor thus emits light only when in close proximity to the donor molecule.

Probe-based detection offers increased specificity and the option of multiplexing reactions, assuming that the assays to be multiplexed are of similar efficiency, can be performed under the same reaction conditions, and the primers and probes used in the assays do not interact with each other or form primer dimers. For multiplexed reactions, each probe should be labeled with a fluorescent reporter with distinctly different emissions spectra. It should be noted that ROX, which is commonly included as a passive reference dye for certain real-time PCR platforms, has a similar emission spectrum as the dye TexasRed and can interfere with detection; therefore, ROX should be omitted from any reaction in which TexasRed, or other probes with similar emissions wavelengths (approximately 600-650 nm), are conjugated to the probe. Additionally, the use of certain fluorescent reporters may be limited by the wavelengths of light which the laboratory's real-time PCR machine, filters, or software can process.

Oligonucleotide probe-based detection methods were initially based upon the annealing and hydrolysis, mediated by *Taq* polymerase, of probes labeled with a fluorescent reporter on one end and a fluorescence quencher on the other (Fig. **1B**). TaqMan® probes are one example of this type of chemistry. As the DNA polymerase extends the template from the primer, in a 5' to 3' direction, it also exhibits 3' exonuclease activity which degrades the probe and releases the fluorescence from the quencher. In these assays, the probe should be designed to anneal to the template at a temperature approximately 10°C higher than that of the primers, to ensure the probe is in place before the primers anneal and the polymerase begins extension. This requirement can produce relatively long probes, especially when detecting GC-poor templates. If the probes are longer than 30-35 bases, quenching may be poor, resulting in high background fluorescence. Minor-groove binding, or MGB, probes were developed to produce shorter probes with the necessary high annealing temperature for hydrolysis assays [38]. In these probes, a moiety that binds to the minor groove of dsDNA is covalently bonded to the 3' end of the probe, in addition to the quencher, providing increased stability of the probe-template duplex and increasing the annealing temperature. LNA probes use modified nucleic acids, "locked nucleic acids", to achieve similar results [39], the higher the number of LNA bases in the probe, the higher the annealing temperature can be for a short oligonucleotide. Both MGB and LNA probes are more specific than traditional hydrolysis probes, and allelic discrimination assays that detect single nucleotide polymorphisms typically use either MGB or LNA technology. Melt temperature analysis cannot be used with any of these hydrolysis probe systems.

In contrast to the hydrolysis probes discussed above, molecular beacon, Scorpion®, and LightCycler (HybProbe) probes depend upon hybridization without hydrolysis. Molecular beacon probes have a stem-loop structure, with a central region that is complementary to the target DNA, flanked on both sides by short, GC-rich sequences that are complementary to each other (Fig. **1C**). Ideally, both the central region of the probe and the GC-rich arms will anneal to their respective targets at a temperature approximately 7-10°C higher than the

PCR primers but lower than the temperature used for the extension phase of the PCR. Similar to hydrolysis probes, molecular beacons are labeled with a fluorescent reporter at the 5' end and a quencher at the 3' end; the hairpin structure of the probe keeps the reporter and quencher in close physical proximity. In the absence of a target, these probes self-anneal and form stable hairpin structures that exhibit no fluorescence. When the target is present, the central portion of the probe hybridizes to the target during the annealing step of PCR, separating the fluorescent reporter and quencher and generating fluorescence. These probes dissociate at the higher temperatures associated with PCR extension and are not hydrolyzed by *Taq* polymerase. After the PCR is complete, melt curve analysis can be used to separate the probe from the template strand; the T_m is slightly decreased in the case of base pair mismatches, providing investigators with some limited capacity to detect polymorphisms within the probe sequence. Because hydrolysis is not required for signal reporting, molecular beacons have also been used to detect DNA and RNA in other applications, including *in situ* hybridization. Careful design of the probe is necessary to establish the appropriate stem-loop structure and ensure that the different portions of the probe anneal at the correct temperatures; in practice, this can make molecular beacon probes more difficult to design than hydrolysis probes. In addition to the dual-labeled molecular beacons that couple a fluorescent reporter with a quencher, there are some chemistries available that use a single fluorophore whose emission is altered by the process of hybridization [40, 41].

Scorpion® probes are essentially a modification of molecular beacons, in which an oligonucleotide probe, with the stem-loop structure, a reporter, and a quencher, is covalently attached to the 5' end of one of the PCR primers (Fig. **1D**). In these systems, the primer is incorporated into the product during amplification; the probe remains attached, and during subsequent cycles, the probe can hybridize to the adjacent end of the product during the annealing step and generate fluorescent signal. As with molecular beacons, the secondary structure of the Scorpion® probe-primer combination is critical. However, use of these probes reduces the number of components that must be optimized to develop the final assay.

Dual-probe hybridization systems (HybProbe, LightCycler®) consist of two separate probes, or a probe and a primer, which anneal on the same strand of the template [4] (Fig. **1E**). The first probe is labeled with a donor fluorophore at the 3' end, and the second oligonucleotide is labeled with an acceptor fluorophore at the 5' end. When the probes are both annealed to the template strand, the fluorophores are separated by a gap of 1-4 nucleotides, allowing fluorescent energy to be transferred from the donor to the acceptor. The use of two probes increases the number of nucleotides used for detection and, thus, the specificity of the assay. As with other hybridization probes, melt curve analysis can be used to assess the strength of probe annealing, with reduced T_m reported in the case of polymorphisms in either probe.

4. EXAMPLES OF REAL-TIME PCR APPLICATIONS

Real-time PCR for the detection of DNA or RNA has found widespread use in both diagnostic and research laboratories. The sensitivity of these assays is superior to conventional PCR and similar to that of nested PCR, but the real-time PCR process is more rapid, quantifiable, and minimizes opportunity for sample cross-contamination. The specificity and ability to multiplex probe-based reactions are also superior to conventional PCR. Finally, quantitative data are invaluable for research studies. Examples of some applications for real-time PCR follow.

4.1. Single-Target Assays

Assays for single targets can be based upon either dye or probe detection of the target and are supported by all real-time PCR platforms. These are simpler to validate than multiplexed assays and can provide sensitive, specific, quantitative detection of a PCR target. To maximize the sensitivity of the assay for rare targets, the efficiency of a single real-time PCR assay should be optimized to be >85%, with no significant primer dimer formation; a detection limit of fewer than 10 gene copies per reaction can be achieved with good design. The quantitative dynamic ranges of these assays typically cover 7-9 orders of magnitude.

Real-time PCR has been extensively adopted for the detection of diverse pathogens, including bacteria, fungi, protozoa, and RNA and DNA viruses. Every type of probe chemistry has been applied to the detection of pathogens, depending upon the equipment of the laboratory developing the assay, the requirements for sensitivity and specificity, and the type of sample and pathogen being detected. Optimization of pathogen-detection assays should emphasize the diagnostic sensitivity and specificity of the assay. Specific concerns include the possibility of inhibition caused by compounds in the nucleic acid sample, a common concern when working with blood, tissue, soil, or fecal specimens, and non-target amplification of host, animal, normal flora, or other background DNA that is also present in the sample.

Real-time PCR has been especially useful for the detection of organisms that are difficult to cultivate, including not-yet-cultivated organisms, viruses, and rickettsial agents (*e.g.*, [35, 42-44]). PCR, including real-time PCR, is also used for the detection and identification of cultivatable but slow-growing pathogens, including *Mycobacterium*, *Histoplasma*, and *Brucella*. Likewise reverse transcriptase real time PCR can detect non-lytic viruses in cell cultures. The high sensitivity of real-time PCR improves the ability to detect DNA that is present at very low levels; for example, several real-time PCR detection chemistries were recently compared for their ability to detect trace amounts of DNA from genetically modified organisms [34]. Because these assays are highly specific, they can also be used for the detection of bacteria in mixed samples in which contaminants might overgrow the agent of interest. The latter approach is especially relevant when looking for a specific pathogen in soil or fecal DNA samples. Finally, PCR detection can be conducted on inactivated samples under biosafety level (BSL) 1 or 2 conditions, a significant consideration when the live pathogen must normally be handled and cultivated under BSL-3, BSL-4, or Select Agent conditions; examples include *Bacillus anthracis* and *Francisella tularensis* [45, 46].

4.2. Multiplex Assays

Multiplexed real-time PCR assays combine several reactions in a single tube, reducing the quantity of reagents needed to screen a sample for multiple targets. Multiplexed assays can be based upon a single primer pair with multiple probes, to discriminate between taxa, or upon multiple primers and probes. The number of colors is typically restricted to three or four, depending upon the capability of the real-time PCR platform and software to be used.

Optimization and validation of multicolor assays is more rigorous, since interactions between the assays can interfere with sensitivity. Typical problems include primer dimers, competition between assays for dNTP's and polymerase, and possible overlap in emissions spectra between reporter fluorophores. The former can be reduced by careful design of the assays, to minimize interactions between all the different primers. Multiplexed assays that include high-copy number targets may require additional dNTP's. Careful selection of the fluorophores to minimize overlap in spectra, while remaining compatible with the equipment and filters available, is also necessary. And, finally, the effects of competition are reduced if all assays included in the multiplex have similar efficiency. This is especially important for assays which use a housekeeping gene to normalize data; similar efficiencies ensure that the housekeeping gene and target gene are amplified at proportional rates.

Multiplexed assays offer distinct advantages when testing for a panel of pathogens, reducing both the number of reactions and the time and labor required to complete testing. When multiplexed reactions replace several conventional or nested PCR, they can also be very cost-effective for screening for panels of pathogens. The transition to multiplexed reactions in a diagnostic virology laboratory, as well as some of the criteria for validation in this setting, was described by Gunson, *et al.*, in 2008 [47].

SNP genotyping, or allelic discrimination, assays are designed to detect and discriminate between two separate alleles at a specific locus. The assays use a single primer set and two probes, each one designed to anneal to one allele of the gene, labeled with different fluorophores; each individual can be identified as either homozygous for one allele, heterozygous, or homozygous for the other allele. Short, high-affinity probes, such as MGB or LNA probes, usually work best for these applications (*e.g.*, [34, 38, 39]).

4.3. Quantitating Gene Copies

Quantitative data have extensive applications in research. These data can be used for experiments that follow the kinetics of experimental or natural infection, to detect changes in the expression of virulence genes, to confirm changes detected using microarrays, and to measure quantitative differences in cytokine mRNA production between individuals or experimental treatment groups [3, 48-50].

Quantitation of gene copies is most commonly achieved by assaying unknown samples at the same time as a standard curve of known, quantified positive control template. Five- or ten-fold dilution series are used to generate the series of standards to be tested; the number of copies in each standard is log-transformed and plotted against the C_T; and then linear regression is used to establish a line of best fit for the standard curve. The quantity of starting copies in each unknown sample is then extrapolated from this equation, based upon the C_T for the sample. The standard caveats for quantitative assays apply: quantitation is valid only within the dynamic range of the assay, and unknown samples must fall within the upper and lower bounds of the standards which were run in the assay. As the field has advanced, several different formulas have been proposed for analyzing quantitative data, with differing assumptions in regards to reaction efficiency [51-55].

To minimize the effect of variations in extraction efficiency, starting template concentration, reverse transcription efficiency, or other sources of variation, quantitative data should be normalized. Quantitation of DNA may be normalized according to the initial volume or mass of the sample which was used for the extraction: gene copies per microliter of blood, copies per milligrams of spleen tissue, copies per colony forming unit (bacteria), *etc.* Data may also be normalized using the concentration of nucleic acids in the extracted sample or the total quantity of DNA included as the template for the real-time PCR.

When RNA is the target, the quantified target is normalized using the transcript levels for a housekeeping gene; the housekeeping gene should have similar levels of mRNA expression in all cells, regardless of infection status or other variables. When housekeeping genes are used to normalize mRNA quantitation, it is advantageous to assay the housekeeping gene in the same reaction as the target gene, to minimize variability in starting template quantity and reverse transcription efficiency. In the most rigorous sense, standard curves are performed for both the target and housekeeping genes, and the absolute number of target gene copies is normalized using the number of housekeeping gene copies. However, normalization does not require standard curves to be used, as long as the efficiencies of the assays used for the target and housekeeping genes are known and are nearly identical. For relative comparisons, the C_T for the RNA target is normalized using the C_T for the housekeeping gene, and these normalized data are compared between individuals or experimental groups.

In practice, relative quantitation of mRNA requires careful consideration both of the mathematical equations and the housekeeping gene(s) used to normalize the data. This field has become highly developed over the last several years, and laboratories wishing to adopt this technology will want to consult recent reviews of this field (*e.g.*, [49, 50, 55-57]). Many real-time PCR platforms have analysis tools built into the software package, and more specialized tools are available, either as free applications or commercial packages [49]. In recent years, the housekeeping genes most commonly used to normalize mRNA data, including β-actin and GAPDH, have been shown to be variable, and therefore unsuitable, under some biological conditions [56, 58]. Selection of an appropriate housekeeping gene is a critical component of this type of study and should be carefully researched in the literature for the tissue type and experimental model.

5. ADOPTING REAL-TIME PCR

Overall, the advantages offered by real-time PCR, including sensitivity, specificity, reduction in sample cross-contamination events, and high-throughput capability, make this technology suitable for use in diagnostic laboratory settings [3, 34, 35, 44, 47]. The benefits of this technology are also closely related to its drawbacks; sensitivity to contamination and inhibition require increased stringency in laboratory cleanliness and work processes. Once a laboratory has experience with real-time PCR processes, additional assays can be added with minimal changes to the standard operating procedures.

Although the design of assays can be technically complex, many assays have already been designed and validated; these are now available in databases dedicated to real-time PCR assays [49, 50] or in published literature. These resources can greatly reduce the initial investment in time and expertise to develop a diagnostic assay for any particular disease. When assays are adopted from other sources, careful attention should be paid to any variables which may not be identical; examples include changing the platform, using a different formulation of master mix, beginning with a different sample type or extraction method, and changing either the template or reaction volume. Even small changes can have significant consequences. For example, the ramp rates for heating and cooling vary between PCR instruments from different manufacturers, and these differences can cause well-validated assays to fail on some platforms. Prior to diagnostic use, each assay should always be tested in-house, to confirm that the sensitivity, specificity, and efficiency are similar to those reported by the original authors. Once this is done, internal controls can be included in every run to provide ongoing quality control data. Further validation is not needed unless one of the variables changes or a run fails the quality control check.

Finally, when an assay is not readily available and must be developed *de novo*, careful planning in the pre-development phase can greatly reduce downstream problems with validation. Selecting the most appropriate nucleic acid extraction method, reverse transcription strategy, and real-time PCR detection chemistry are important steps in the planning process. A careful review of the literature can also help identify problems, and solutions, that have been reported by other laboratories. All of these steps reduce the barriers to effectively incorporate real-time PCR into a diagnostic laboratory setting.

REFERENCES

[1] Higuchi R, Fockler C, Dollinger G, Watson R. Kinetic PCR analysis: real-time monitoring of DNA amplification reactions. Biotechnology (N Y) 1993; 11(9): 1026-30.
[2] Gudnason H, Dufva M, Bang DD, Wolff A. Comparison of multiple DNA dyes for real-time PCR: effects of dye concentration and sequence composition on DNA amplification and melting temperature. Nucleic Acids Res 2007; 35(19): e127, 8pp.
[3] Kaltenboeck B, Wang C. Advances in real-time PCR: application to clinical laboratory diagnostics. Adv Clin Chem 2005; 40: 219-59.
[4] Lyon E. Mutation detection using fluorescent hybridization probes and melting curve analysis. Expert Rev Mol Diagn 2001; 1(1): 92-101.
[5] Heid CA, Stevens J, Livak KJ, Williams PM. Real time quantitative PCR. Genome Res 1996; 6(10): 986-94.
[6] Wilson WC, Hindson BJ, O'Hearn ES, *et al.* A multiplex real-time reverse transcription polymerase chain reaction assay for detection and differentiation of Bluetongue virus and Epizootic hemorrhagic disease virus serogroups. J Vet Diagn Invest 2009; 21(6): 760-70.
[7] Aslanzadeh J. Preventing PCR amplification carryover contamination in a clinical laboratory. Ann Clin Lab Sci 2004; 34(4): 389-96.
[8] Al-Soud WA, Radstrom P. Purification and characterization of PCR-inhibitory components in blood cells. J Clin Microbiol 2001; 39(2): 485-93.
[9] Wilson IG. Inhibition and facilitation of nucleic acid amplification. Appl Environ Microbiol 1997; 63(10): 3741-51.
[10] Cler L, Bu D, Lewis C, Euhus D. A comparison of five methods for extracting DNA from paucicellular clinical samples. Mol Cell Probes 2006; 20(3-4): 191-6.
[11] Desloire S, Valiente MC, Chauve C, Zenner L. Comparison of four methods of extracting DNA from *D. gallinae* (Acari: Dermanyssidae). Vet Res 2006; 37(5): 725-32.
[12] Dundas N, Leos NK, Mitui M, Revell P, Rogers BB. Comparison of automated nucleic acid extraction methods with manual extraction. J Mol Diagn 2008; 10(4): 311-6.
[13] Exner MM, Lewinski MA. Isolation and detection of *Borrelia burgdorferi* DNA from cerebral spinal fluid, synovial fluid, blood, urine, and ticks using the Roche MagNA Pure system and real-time PCR. Diagn Microbiol Infect Dis 2003; 46(4): 235-40.
[14] Metwally L, Fairley DJ, Coyle PV, *et al.* Improving molecular detection of *Candida* DNA in whole blood: comparison of seven fungal DNA extraction protocols using real-time PCR. J Med Microbiol 2008; 57(Pt 3): 296-303.
[15] Moriarity JR, Loftis AD, Dasch GA. High-throughput molecular testing of ticks using a liquid-handling robot. J Med Entomol 2005; 42(6): 1063-7.
[16] Smith K, Diggle MA, Clarke SC. Comparison of commercial DNA extraction kits for extraction of bacterial genomic DNA from whole-blood samples. J Clin Microbiol 2003; 41(6): 2440-3.
[17] Tomaso H, Kattar M, Eickhoff M, *et al.* Comparison of commercial DNA preparation kits for the detection of Brucellae in tissue using quantitative real-time PCR. BMC Infect Dis 2010; 10: 100, 5p.

[18]   McAvin JC, Bowles DE, Swaby JA, *et al.* Identification of *Aedes aegypti* and its respective life stages by real-time polymerase chain reaction. Mil Med 2005; 170(12): 1060-5.

[19]   Butler JM. DNA extraction from forensic samples using chelex. Cold Spring Harb Protoc 2009; doi:10.1101/pdb.prot5229.

[20]   Giraffa G, Rossetti L, Neviani E. An evaluation of chelex-based DNA purification protocols for the typing of lactic acid bacteria. J Microbiol Methods 2000; 42(2): 175-84.

[21]   Mygind T, Birkelund S, Birkebaek NH, Ostergaard L, Jensen JS, Christiansen G. Determination of PCR efficiency in chelex-100 purified clinical samples and comparison of real-time quantitative PCR and conventional PCR for detection of *Chlamydia pneumoniae*. BMC Microbiol 2002; 2: 17, 8pp.

[22]   Tani H, Tada Y, Sasai K, Baba E. Improvement of DNA extraction method for dried blood spots and comparison of four PCR methods for detection of *Babesia gibsoni* (Asian genotype) infection in canine blood samples. J Vet Med Sci 2008; 70(5): 461-7.

[23]   Poli F, Cattaneo R, Crespiatico L, Nocco A, Sirchia G. A rapid and simple method for reversing the inhibitory effect of heparin on PCR for HLA class II typing. PCR Methods Appl 1993; 2(4): 356-8.

[24]   Peirson SN, Butler JN. RNA extraction from mammalian tissues. Methods Mol Biol 2007; 362: 315-27.

[25]   Blow JA, Mores CN, Dyer J, Dohm DJ. Viral nucleic acid stabilization by RNA extraction reagent. J Virol Methods 2008; 150(1-2): 41-4.

[26]   Turell MJ, Spring AR, Miller MK, Cannon CE. Effect of holding conditions on the detection of West Nile viral RNA by reverse transcriptase-polymerase chain reaction from mosquito (Diptera: Culicidae) pools. J Med Entomol 2002; 39(1): 1-3.

[27]   Thomas FC. Comparison of some storage and isolation methods to recover bluetongue virus from bovine blood. Can J Comp Med 1984; 48(1): 108-10.

[28]   Deng MY, Wang H, Ward GB, Beckham TR, McKenna TS. Comparison of six RNA extraction methods for the detection of classical swine fever virus by real-time and conventional reverse transcription-PCR. J Vet Diagn Invest 2005; 17(6): 574-8.

[29]   Rump LV, Asamoah B, Gonzalez-Escalona N. Comparison of commercial RNA extraction kits for preparation of DNA-free total RNA from *Salmonella* cells. BMC Res Notes 2010; 3: 211, 5pp.

[30]   Gallo RC. Reverse transcriptase, the DNA polymerase of oncogenic RNA viruses. Nature 1971; 234(5326): 194-8.

[31]   Kato CY, Mayer RT. An improved, high-throughput method for detection of bluetongue virus RNA in *Culicoides* midges utilizing infrared-dye-labeled primers for reverse transcriptase PCR. J Virol Methods 2007; 140(1-2): 140-7.

[32]   Gelderblom HR. Structure and Classification of Viruses. In: Baron S, editor. Medical Microbiology. 4th ed. Galveston, TX: University of Texas Medical Branch at Galveston; 1996.

[33]   Arya M, Shergill IS, Williamson M, Gommersall L, Arya N, Patel HR. Basic principles of real-time quantitative PCR. Expert Rev Mol Diagn 2005; 5(2): 209-19.

[34]   Buh GM, Tengs T, La Paz JL, *et al.* Comparison of nine different real-time PCR chemistries for qualitative and quantitative applications in GMO detection. Anal Bioanal Chem 2010; 396(6): 2023-9.

[35]   Hoffmann B, Beer M, Reid SM, *et al.* A review of RT-PCR technologies used in veterinary virology and disease control: sensitive and specific diagnosis of five livestock diseases notifiable to the World Organisation for Animal Health. Vet Microbiol 2009; 139(1-2): 1-23.

[36]   McChlery SM, Clarke SC. The use of hydrolysis and hairpin probes in real-time PCR. Mol Biotechnol 2003; 25(3): 267-74.

[37]   Zhang T, Fang HH. Applications of real-time polymerase chain reaction for quantification of microorganisms in environmental samples. Appl Microbiol Biotechnol 2006; 70(3): 281-9.

[38]   Walburger DK, Afonina IA, Wydro R. An improved real time PCR method for simultaneous detection of C282Y and H63D mutations in the HFE gene associated with hereditary hemochromatosis. Mutat Res 2001; 432(3-4): 69-78.

[39]   Latorra D, Campbell K, Wolter A, Hurley JM. Enhanced allele-specific PCR discrimination in SNP genotyping using 3' locked nucleic acid (LNA) primers. Hum Mutat 2003; 22(1): 79-85.

[40]   Vaughn CP, Elenitoba-Johnson KS. Hybridization-induced dequenching of fluorescein-labeled oligonucleotides: a novel strategy for PCR detection and genotyping. Am J Pathol 2003; 163(1): 29-35.

[41]   Venkatesan N, Seo YJ, Kim BH. Quencher-free molecular beacons: a new strategy in fluorescence based nucleic acid analysis. Chem Soc Rev 2008; 37(4): 648-63.

[42]   Doyle CK, Labruna MB, Breitschwerdt EB, *et al.* Detection of medically important *Ehrlichia* by quantitative multicolor TaqMan real-time polymerase chain reaction of the *dsb* gene. J Mol Diagn 2005; 7(4): 504-10.

[43]   Fenollar F, Raoult D. Molecular diagnosis of bloodstream infections caused by non-cultivable bacteria. Int J Antimicrob Agents 2007; 30 Suppl 1: S7-15.

[44]   Pusterla N, Madigan JE, Leutenegger CM. Real-time polymerase chain reaction: a novel molecular diagnostic tool for equine infectious diseases. J Vet Intern Med 2006; 20(1): 3-12.

[45]	Jones SW, Dobson ME, Francesconi SC, Schoske R, Crawford R. DNA assays for detection, identification, and individualization of select agent microorganisms. Croat Med J 2005; 46(4): 522-9.

[46]	Versage JL, Severin DD, Chu MC, Petersen JM. Development of a multitarget real-time TaqMan PCR assay for enhanced detection of *Francisella tularensis* in complex specimens. J Clin Microbiol 2003; 41(12): 5492-9.

[47]	Gunson RN, Bennett S, Maclean A, Carman WF. Using multiplex real time PCR in order to streamline a routine diagnostic service. J Clin Virol 2008; 43(4): 372-5.

[48]	Giulietti A, Overbergh L, Valckx D, *et al.* An overview of real-time quantitative PCR: applications to quantify cytokine gene expression. Methods 2001; 25(4): 386-401.

[49]	Thellin O, ElMoualij B, Heinen E, Zorzi W. A decade of improvements in quantification of gene expression and internal standard selection. Biotechnol Adv 2009; 27(4): 323-33.

[50]	VanGuilder HD, Vrana KE, Freeman WM. Twenty-five years of quantitative PCR for gene expression analysis. Biotechniques 2008; 44(5): 619-26.

[51]	Boggy GJ, Woolf PJ. A mechanistic model of PCR for accurate quantification of quantitative PCR data. PLoS One 2010; 5(8): e12355.

[52]	Gallup JM, Ackermann MR. The 'PREXCEL-Q Method' for qPCR. Int J Biomed Sci 2008; 4(4): 273-93.

[53]	Pfaffl MW. A new mathematical model for relative quantification in real-time RT-PCR. Nucleic Acids Res 2001; 29(9): e45.

[54]	Rutledge RG, Cote C. Mathematics of quantitative kinetic PCR and the application of standard curves. Nucleic Acids Res 2003; 31(16): e93.

[55]	Schefe JH, Lehmann KE, Buschmann IR, Unger T, Funke-Kaiser H. Quantitative real-time RT-PCR data analysis: current concepts and the novel "gene expression's CT difference" formula. J Mol Med 2006; 84(11): 901-10.

[56]	Guenin S, Mauriat M, Pelloux J, Van WO, Bellini C, Gutierrez L. Normalization of qRT-PCR data: the necessity of adopting a systematic, experimental conditions-specific, validation of references. J Exp Bot 2009; 60(2): 487-93.

[57]	Huggett J, Dheda K, Bustin S, Zumla A. Real-time RT-PCR normalisation; strategies and considerations. Genes Immun 2005; 6(4): 279-84.

[58]	de Jonge HJ, Fehrmann RS, de Bont ES, *et al.* Evidence based selection of housekeeping genes. PLoS One 2007; 2(9): e898.

CHAPTER 2

Design and Optimization of Real-Time PCR Assays

Raymaekers Marijke[*]

Clinical Laboratory, Jessa Hospital, Site Virga Jesse, Stadsomvaart 11, 3500 Hasselt, Belgium

Abstract: With increased use of real-time polymerase chain reaction technology in molecular diagnostics, consistent procedures for design, optimization and validation of molecular diagnostic methods are needed. This chapter describes a practical guiding principle that can be used in different steps of the design and validation of in-house developed real-time PCR assays. The use of the described guidelines leads to more efficient and standardized optimization and validation. Ultimately, this results in a reliable and robust molecular diagnostic assay. A statistical follow-up of the performance of the assay is included and can be achieved by determination of target values and reproducibility of internal quality controls. Since this guiding principle is independent of environment, equipment and specific applications, it can be used in any laboratory.

Keywords: Oligonucleotide design; in-house assay; target gene; optimization; validation; specificity; melting curve analysis; amplicon sequencing; linearity; annealing temperature; internal quality control.

1. INTRODUCTION

Shortly following the invention of the polymerase chain reaction (PCR) by Saiki *et al.* [1] and Mullis *et al.* [2], PCR kinetics could be analyzed by Higuchi *et al.* [3, 4]. The described system made the detection of accumulating PCR products possible, by using an intercalating dye ("real-time PCR"). In 1991, Holland *et al.* [5] demonstrated the cleavage of a target-specific probe during PCR, using the 5' nuclease activity of *Taq* DNA polymerase.

Real-time PCR was further improved by the development of fluorogenic probes [6, 7] that enabled monitoring through the production of a fluorescent signal. The latter is generated only in the case of specific hybridization between probe and target.

Compared to conventional PCR, real-time PCR methodology allows the non-laborious, reliable detection and quantification of most nucleic acid target sequences, in addition to detection and differentiation. Thanks to these features, real-time PCR has revolutionized molecular biology and an extensive number of applications have been developed, both in research and in clinical diagnostics [8, 9]. The majority of these applications are non-commercial in-house developed assays. As standardization and quality assurance of molecular diagnostic methods is required, guidelines for design, optimization and validation of real-time PCR assays can be useful tools, in addition to procedures on good laboratory practice or on subdivisions of the validation process [10-14]. This chapter describes a guideline that can be used for different steps of the development and validation process of real-time PCR assays. This guiding principle leads to efficient and standardized optimization and validation. Because it is independent of reaction environment, equipment and specific applications, it can be exchanged between laboratories.

2. DESIGN, OPTIMIZATION AND VALIDATION OF REAL-TIME PCR ASSAYS: A GUIDING PRINCIPLE

The guidelines described in Evidence Based Laboratory Medicine (EBLM) [15] and the critically appraised topic (CAT) method [16] are valuable tools in the selection of an examination procedure. These can be

*Address correspondence to Raymaekers Marijke:** Scientific Collaborator, Clinical Lab, Jessa Hospital, site Virga Jesse, Stadsomvaart 11, 3500 Hasselt, Belgium; Tel: +32 11 30 97 02; Fax: +32 11 30 97 50: E-mail: marijke.raymaekers@jessazh.be

equally applied to molecular diagnostic tests. This chapter focuses only on the first two steps of a CAT: the technical and diagnostic performance of molecular diagnostic assays. It describes a guiding principle that is based on literature, existing guidelines and personal experience. It includes general recommendations and can be used for all real-time PCR assays. The complete list is depicted in Table **1** [17] and is clarified below. Test-specific criteria should be defined for each individual test before the collection of validation data starts. The postulated aims should at least describe the clinical purpose (patient and sample type), the technique used, the target sequence and the expected clinical and technical performance of the assay.

2.1. Design

2.1.1. In-House Assay or Commercial Assay

For many clinically important parameters, few commercial real-time PCR assays are yet available and it is often necessary to develop an in-house assay. The checklist described in Table **1** can be used as a template to guide the validation process. The exact interpretation depends on the platform or assay under validation. Some of the items might not be applicable and are preferentially recorded in the report of validation as "not applicable", indicating that these items were taken into account [10, 18].

Table 1: Checklist for design, optimization and validation of real-time PCR assays

1. Design
- Commercial assay or in-house assay
- Choice of target gene
- Choice of detection method
- Choice of chemical components, choice of oligonucleotides
  - Tm of primers (*e.g.*, 58-60 °C)
  - GC content of oligonucleotides: 30-70 %
  - Not more than 2 C or G in last 5 positions at 3' end of primer
  - Length of amplicon: max 400 bp
  - No more than 4 constitutive guanines
  - Avoid primer-dimer formation
  - Length of primer: 18-24 base pairs
  - T_m of probe (*e.g.*, 68-70 °C)
  - More C than G in probe
- Verification of oligonucleotide specificity: Expectation value ≤ 0.01
- Choice of sample material and sample processing
- Quantification strategies
  - Standard curve method
  - Comparative method
- Normalization

2. Optimization of reaction conditions
- Optimization of primer and probe concentration
- Optimization of annealing temperature
- Optimization of template nucleic acid concentration

3. Validation
- Verification of amplification
  - Melt curve analysis
  - Gel electrophoresis
  - Amplicon sequencing + BLAST
- PCR characteristics
  - slope *m*: Ct = log conc × *m* + y-intercept (criteria : - 3.6 ≤ m ≤ - 3.1)
  - Efficiency E: E = $10^{-1/slope} - 1$ (criteria : 0.9 ≤ E ≤ 1.1)
  - Coefficient of correlation r^2 (criteria : 0.99 ≤ r² ≤ 0.999)

- Analytical verification
  - ➤ Precision
  - ➤ Trueness (accuracy)
  - ➤ Linearity, measuring range
  - ➤ Limit of detection (COI ≥95 %) /analytical sensitivity
  - ➤ Limit of quantification
  - ➤ Analytical specificity
- Clinical verification
  - ➤ Clinical question (CAT)
  - ➤ Clinical performance
  - ➤ Correlation to disease or disorder
    - ▪ Negative predictive value
    - ▪ Positive predictive value
  - ➤ Comparison to current methods / standards
- Internal Quality Control
  - ➤ Amplification and inhibition control
  - ➤ Negative control
  - ➤ Statistical follow-up of a positive control
  - ➤ Revision of oligonucleotide sequences
- Proficiency testing

2.1.2. Choice of the Target Gene

The choice of an appropriate nucleic acid target is the first step in the development of an in-house assay. A literature review often reveals which target is most suitable for a particular assay. A specific and conserved nucleic acid target sequence may be selected for the detection of defined taxa such as viruses, bacteria, parasites, fungi and protozoa. To increase sensitivity, a target sequence that is repeated within the genome and shows high copy numbers can be chosen.

In real-time PCR assays used for the detection of somatic mutations, rearrangements, breakpoint fusion regions of chromosome aberrations, fusion-gene transcripts, aberrant genes, and aberrantly expressed genes, the region of aberration should be targeted. Additionally, for reverse transcription (RT) real-time PCR assays with haematological applications, primers and probes should span an exon-exon splice junction, enabling amplification and detection of RNA sequences only. This prevents co-amplification of genomic DNA, which can compromise the sensitivity and efficiency of the assay by competition between the desired PCR product and the product derived from genomic DNA. Screening genome databases with the amplicon sequence helps to ensure that an assay does not detect pseudogenes [19].

In general, G + C rich regions (greater than 60%) in the target sequence should be avoided because they are difficult to amplify [20]. Stretches of polypurines or polypyrimidines (longer than four stretches) within the expected amplicons should also be avoided [21].

2.1.3. Choice of the Detection Method

Real-time PCR detects DNA amplification by monitoring increases in fluorescence. Fluorescent reporters can be nonspecific labels and sequence-specific probes. An intercalating dye gives the opportunity to detect non-sequence-specific amplified products. Melting curve analysis allows for extrapolated examination of the dsDNA levels. However, mis-priming events can generate a false positive-signal. The first dye used as a DNA-binding fluorophore was ethidium bromide [3, 4]. More recently, a less toxic SYBR® Green dye and its derivatives have become widely accepted, and alternative dyes are proposed. The SYBR® Green dye has a number of limitations such as inhibition of the PCR, preferential binding to GC-rich sequences and effect on melting curve analysis [22-24]. A study by Gudnason *et al.* [25], investigating the inhibitory effect on the PCR, the effect on DNA melting temperature and possible binding to GC-rich sequences for 15 different intercalating dyes, showed that the use of SYTO-82 and SYTO-13 will simplify the development of multiplex assays and increase the sensitivity of real-time PCR.

Specific hybridization occurs between a fluorogenic probe and target DNA, and probes can be labeled with two kinds of dye: (i) fluorophores with intrinsically strong fluorescence, which are brought in contact with a quencher molecule through structural design and (ii) fluorophores that can change their fluorescence capacities upon binding the target DNA. Examples of the former kind of probes are hydrolysis probes (based on oligonucleotides [6, 7] or on locked nucleic acids (LNA) [26, 27]), minor groove binding probes [28], molecular beacons [29, 30], and hybridization probes (also called FRET) [31]. More recently, fluorescence-labelled primers were developed [32]. The second kind of probe includes light-up probes [33] and displacement probes [34]. The advantages and disadvantages of different chemistries have been discussed in several publications [11, 35-42] (Table **2**).

Table 2: Comparison of the various technologies available for real-time PCR (adapted from Gunson *et al.*[11])

Chemistry	Advantages	Disadvantages
Hydrolysis probe	Specific, less probe mismatch, increased fluorescence	Probe mismatch can lead to false negatives
Hydrolysis LNA probe	Increased thermal stability, use of shorter probes, high specific/reproducible, for multiplex assays	Expensive
Minor groove binding probe	Specific, allows use of short oligoprobes, ideal for SNP	Susceptible to mismatch, few suppliers
Molecular beacons	Very specific (hairpin loop)	Increased tendency for probe mismatch, reduced fluorescence
FRET	Can detect single nucleotide differences, exact match to DNA	Requires strict optimization of probe design and accurate thermal denaturation profile
Labelled primers	Cheap, sensitivity comparable with probe based assay, less homology needed	Primer-dimer formation, strict design criteria

A new emerging technology is high resolution melt curve analysis (HRM) [43, 44], which characterizes nucleic acids based on their dissociation behaviour. It combines the principle of intercalating dyes, melt curve analysis and specific statistical analysis. The most important applications are single nucleotide polymorphism (SNP) typing, gene scanning, DNA mapping, DNA fingerprinting and DNA methylation analysis [45, 46].

2.1.4. Choice of Chemical Components

A well designed robust real-time PCR assay consists of an optimized mixture of reagents and thermal reaction parameters. The basic components of a real-time PCR mixture are: a thermostable DNA polymerase, oligonucleotide primers, intercalating dye or a fluorescently labeled oligonucleotide probe, nucleotides, $MgCl_2$, KCl and Tris-HCl [47-50]. Often uracil-N-glycosylase (UNG) is added to remove contaminating amplified material [51], although a report by Espy *et al.* [52] seems to indicate that UNG treatment did not inactivate amplicons with a length smaller or equal to 100 base pairs (bp). UNG has also been described to be incompletely deactivated at elevated temperatures [53].

Inactivity of commercial hot start *Taq* DNA polymerases prevents non-specific amplification at room temperature. Inactivation is typically achieved through an antibody directed against the active site of the enzyme [54] or by means of a recombinant *Taq* DNA polymerase with a sequence deletion [55]. The thermostable DNA polymerase is activated after a first heating step. This hot-start strategy is widely used in real-time PCR. Commercially available reaction master mixes, containing nucleotides, $MgCl_2$, KCl, Tris-HCl, and a thermostable DNA polymerase, can reduce labour and potential for human error in preparation of reaction mixtures. When using these master mixes, generally only the oligonucleotide concentrations will need optimization. Mixtures containing UNG or intercalating dye are also available.

Designing the optimal primer and probe sequences are crucial for a successful and reliable PCR. For the design of primers and probes, effective criteria are described in literature (Table **1**) [56, 57]. Several software packages, such as Primer Express [56], Primer 3 [58], Oligo [59] and VECTOR NTI, are currently available to design sets of primers and probe. However, it should be confirmed that the suggested primers and probe set meet the criteria listed in Table **1**.

The melting temperature (T_m) of the oligonucleotides is an indicator of the hybridization strength of oligonucleotides. Although many attempts have been made to predict the T_m [60-62], the formula that calculates the T_m most accurately is based on the nearest-neighbor model. In this model, thermodynamic values for hybridization depend on interactions between a particular base and its nearest neighbors [62-65].

Guidelines concerning the T_m of primers used in two-step protocols suggest annealing and extension at 60 °C [56]. In most three-step PCR protocols, the primer annealing may be within the range of 50 to 60 °C and primer extension is performed at 72 °C, which is the optimal temperature for the *Taq* polymerase. However, hydrolysis probes elongate at 60 °C, which was demonstrated to be equally efficient [5]. The T_m of sense and antisense primers should be similar (max difference of 2°C) to ensure simultaneous hybridization and elongation [64].

Primer-template hybrids are stabilized when the *Taq* polymerase extends the primer. The fluorogenic probe is not extended and the probe-template hybrid must be stabilized using a probe with a higher T_m than the primer-template hybrids and higher than the annealing temperature. To ensure a strong binding of the probe during the annealing phase, hydrolysis and hybridization probes should have a T_m that is 5-10°C higher than the T_m of the primers [5, 66].

Oligonucleotides with high G + C content will stabilize probe hybridization. The presence of G or C within the last 5 bases from the 3' end of the primers (GC clamp) helps promote specific binding at the 3' end due to stronger bonding of G and C. However, a high G + C content at the 3' end (more than 2 C or G in the last 5 positions) of a primer may prevent the complete annealing of the remainder of the primer sequence and reduce the specificity of the reaction [67].

Short amplicons (less than 400 bp) are more efficiently amplified and have less potential for secondary structure [68]. It is presumed that elevated primer extension temperatures are required to denature any secondary structures that may develop in the template and may block extension [35]. The constitutive presence of guanines may fold the template into a tetraplex structure, which is very stable and cannot be transcribed by the polymerase [69]. Self-complementary regions in the template can fold into hairpin and other structures that interfere with extension and reduce the sensitivity of the assay. Consequently, real-time PCR amplicons should be short with low capacity to fold. The software programme designed by Zuker *et al.* [70] can be a useful tool for the prediction of secondary structures.

Primers and probes also should have a low potential to form secondary structures, including self- and cross-hybridization with other oligonucleotides in the PCR ("primer-dimer") [21, 35, 71].

There is no consensus on the size of the primers, but generally primers ranging between 18 and 24 nucleotides are used. Shorter primers (< 17 bases) decrease specificity [64], while longer primers (> 30 bases) are not necessarily more specific [72] and the T_m calculations become less reliable. The design of optimal probes should focus on their hybridization specificity rather than their length. Longer probes allow more mismatches and do not improve the efficiency. Shorter probes increase the chance of nonspecific appearance of these sequences in test material (lower specificity) but exhibit a higher penalty on mismatches. Probes should contain higher C than G contents because such probes produce a greater normalized change in fluorescence (ΔRn). A larger ΔRn allows easier interpretation of the results because positive signals are more easily differentiated from background signals [11, 56].

2.1.5. Verification of Oligonucleotide Design

Once the oligonucleotide sequences are selected, the specificity of the primers and probe should be verified by using the BLAST algorithm [73, 74]. Primers and probe should be verified together, not separately. This

BLAST algorithm performs sequence-similarity searches against various databases, returning a set of gapped alignments with links to full database records. The query coverage and the maximum identity should be 100 % unless the primers contain degeneracy. Each alignment returned by BLAST is scored and assigned a measure of statistical significance as the expectation (E) value, which is an indicator of the probability for finding the match by chance. The E-value is a widely accepted measure of the potential biological relationship. Lower E-values represent higher likelihood of having an underlying biological relationship. Sequences with E-values equal to or smaller than 0.01 are most often found to be homologous [75, 76].

2.1.6. Choice of Sample Material and Sample Processing

The choice of a disease-specific sample must be based on available scientific evidence such as peer-reviewed articles, guidelines and expert opinions. The sample type and patient population for which the test will be designed has to be well defined because it might influence other test selection criteria. The results obtained with a certain method for a given population may not be comparable for another population. Analysis of different sample types within the same population may also give different results [77].

The procedure for sample collection, transport and storage has significant influence on the concentration and stability of the nucleic acids of the test sample. The stability of each sample type can be validated by peer-reviewed reference articles, recommendations of the assay manufacturer or by investigator validation. So far, only small studies on the stability and storage conditions of primary samples have been published [9, 78-83]. The clinical laboratory of standards institute (CLSI) published a guideline [84] describing pre-examination procedures.

A sample material specific for the validation approach is absolutely necessary to account for possible matrix-induced effects [85]. The performance of a diagnostic PCR may be limited by the presence of inhibitory substances within individual samples. PCR inhibitors typically interfere with the action of the DNA polymerase [86], but can also degrade nucleic acids or interfere with the cell lysis procedure [87]. Therefore, efficient sample processing procedures prior to PCR are needed to maintain the performance of the test. Correct sample processing should remove PCR inhibitors, concentrate the target nucleic acids, and turn a heterogeneous biological sample into a homogeneous PCR-compatible sample [85, 88, 89]. Numerous kits for extraction of nucleic acids are commercially available and there is a tendency towards automated extraction. It is advised to validate the recovery of nucleic acids, removal of PCR inhibitors and compatibility of nucleic acid storage buffers with the real-time PCR assay [50].

2.1.7. Quantification Strategies

Quantification of the DNA and RNA targets with real-time PCR can be performed by running standard curves or by the comparative method [49]. The first method depends upon a linear relationship between the input copy number and the increase of fluorescence in the exponential phase. Quantification can be either absolute or relative. Absolute quantification requires the construction of a standard curve, plotting the threshold cycle (Ct) values against the algorithm of the input copy numbers of standards with known concentrations. Standard material must be stable, reliable, and precisely quantified. The copy numbers can be calculated after linear regression of the standard curve. Absolute quantification allows the exact determination of copy number per cell, per total RNA/DNA concentration, or per volume of sample matrix. Relative quantification determines the changes of steady-state transcription of a gene. A relative standard curve consists of a dilution series created using a calibrator with arbitrary units.

To circumvent the use of standard material and standard curves, relative changes in the expression of the target gene can also be determined by the use of the comparative ΔΔCt when PCR efficiencies are the same [90], or by the mathematical model proposed by Pfaffl when PCR efficiencies are different [91, 92]. To compensate for differences in the amount of biological material in the tested sample, normalization is necessary. Many normalization procedures have been suggested but the most popular strategy is normalization to internal reference genes [35]. Finding appropriate reference genes for data normalization is problematic as evidence suggests that there is no universal reference gene with a constant expression in all tissue types [49, 93-95].

2.2. Optimization of Reaction Conditions

Optimization of reaction conditions can increase the efficiency and specificity of the amplification process. The composition of the reaction mixture and the thermal cycling profile should be optimized for only one parameter at a time.

2.2.1. Optimization of Primer and Probe Concentration

Optimization of both primer and probe concentrations limits the primer-dimer formation and ensures the most sensitive and most efficient assay. An optimization matrix should be applied to the primers and the probe using a starting template at a concentration that is expected to yield reproducible results and a no template control (NTC). The matrix consists of a range of forward and reverse primer concentrations (such as 50, 300 and 900 nM), and a range of probe concentrations (such as 50, 100, 150 and 200 nM), resulting in multiple combinations (36 in the example ranges given). The optimal primer and probe concentrations are those for which the lowest Ct value and the highest ΔRn are obtained for the template and no amplification for the NTC [56]. In reality, more than one combination will respond to these criteria and additional testing should be performed for these combinations, using two tenfold dilutions of a positive control near the expected limit of detection. The optimal primer and probe concentrations are those for which the Ct value and the ΔRn fulfill the criteria listed above and a difference in Ct values between the two dilutions of approximately three is obtained [56]. For multiplex PCR assays, optimization reactions should be performed first in a reaction with a single set of primers and probe(s).

2.2.2. Optimization of Annealing Temperature

Although the design of the primers and probes assumes that the annealing will be performed at 60°C (or other input temperature if a three stage reaction is designed), software programmes do not account for the stabilizing effect of the DNA polymerase, which makes optimization of the annealing temperature necessary. PCR characteristics are defined for a standard curve after performing the real-time PCR assay at a range of annealing temperatures (such as 58°C, 60°C and 62°C). The temperature for which the PCR characteristics meet the criteria listed above is the optimal temperature.

2.2.3. Optimization of Sample Input

The DNA or cDNA input must be optimized to ensure a maximum sensitivity with minimum inhibition. PCR characteristics can be defined for standard curves made with different amounts of template (*e.g.*, 2.5, 5, 7.5, 10 and 12.5 µl) added to the reaction mixture. The amount of template for which the PCR characteristics meet the criteria listed above is the optimal amount.

2.3. Validation

For the validation of real-time PCR assays, reference materials (*e.g.*, cell lines, proficiency panels, panels from commercial companies and standards from NIBSC or ATCC) should be used. Reference materials can be used only if a clear statement concerning the content and the concentration is provided. Clinical samples can be used for validation only if they are characterized by a second method.

2.3.1. Verification of Amplification

The absence of primer-dimer formation should be confirmed with a well-documented sample through melting-curve analysis, resulting in a single peak. The length of the amplicon, determined with gel electrophoresis, should be of the expected size. Sequence analysis of the amplification product should be compared with target sequences from GenBank [73].

2.3.2. PCR Characteristics

PCR characteristics can be defined from a standard curve based on tenfold serial dilutions of the DNA or cDNA, within the dynamic range of the method (Fig. **1**). Each dilution should be analyzed in triplicates. In our experience, reliability of results is assumed only when the standard curve is analyzed ten times on different days. Ct values of diluted reference materials are plotted against the logarithm of the samples' concentrations, and the number of template copies or dilution factor [91, 96]. The slope (*m*) must be calculated by linear regression. For *m* to be an indicator of real amplification with exponentially increasing fluorescence (rather than signal drift), there has to be a breakpoint in the amplification plot (Fig. **2**). Signal drift can be produced for a number of reasons. True positive samples may show signal drift because of sub-optimal PCR conditions, inhibition and primer mismatches. Negative samples can show signal drift due to the breakdown of the probe, which occurs in the later cycles if the cycle number is high.

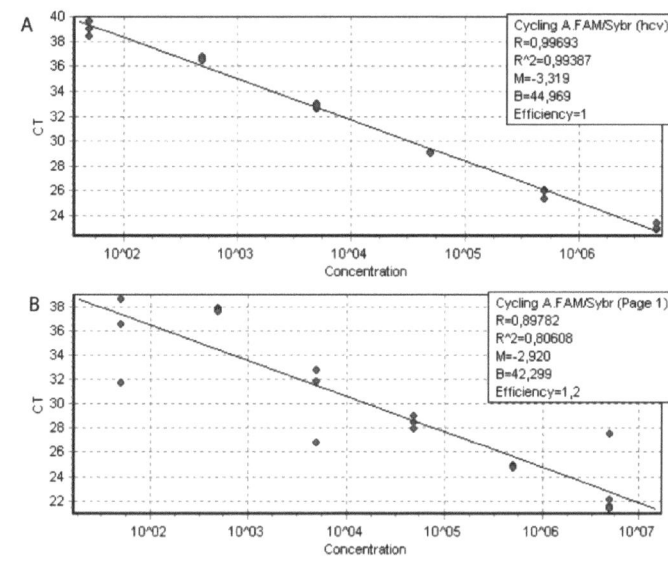

Figure 1: Examples of standard curves based on tenfold serial dilutions of cDNA. (A): the standard curve has characteristics that fulfil the listed criteria; (B): standard curve has characteristics that do not fulfil the listed criteria.

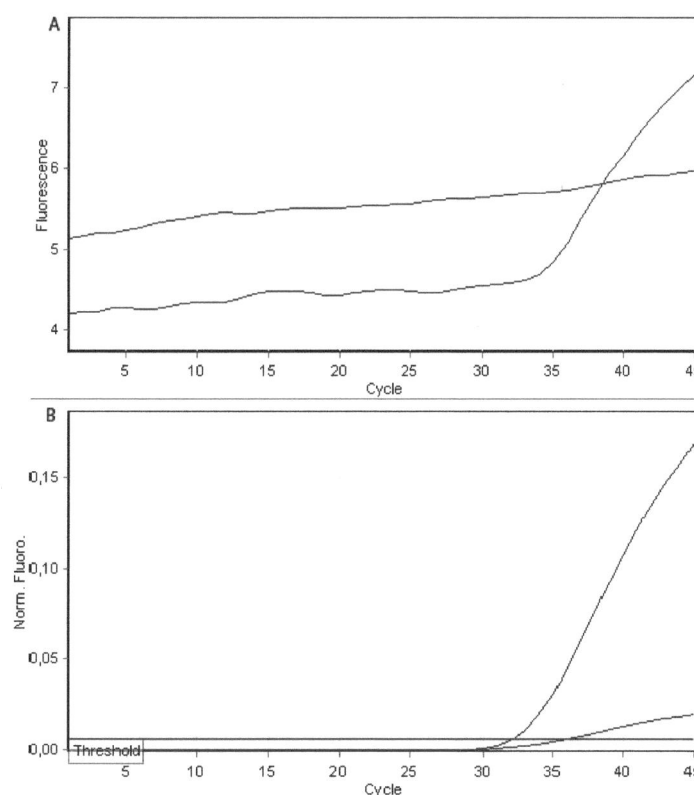

Figure 2: Example of a negative sample showing signal drift. Analysis of raw data shows that there is no increase of fluorescence (A) compared to a real positive sample (blue). This negative sample (purple) shows an increase in fluorescence when analysed with the quantification option (B), showing a weak positive result.

The slope of the linear regression line, ideally - 3.3219, results in a real-time PCR efficiency (*E*) of 1. At a PCR efficiency of 1, the number of target molecules exactly doubles in one PCR cycle [38]. Slopes between - 3.1 and - 3.6, with efficiency between 90 and 100 % are generally acceptable. A number of variables (*e.g.*, PCR inhibitors, PCR enhancers, DNA degradation, DNA concentration, length of the amplicon, secondary structure, and primer quality) can affect the efficiency of the PCR [35, 97, 98].

The efficiency, calculated by the standard curve method, assumes *E* is equal between quantification standards and unknown test samples. The sample-specific amplification efficiency can be calculated *via* sigmoidal [99-102] or logistic [103] curve fitting, [104] and is theoretically 2 [105]. The coefficient of correlation (r^2) is a measure of the relationship between two variables, or more specifically, of their linear relationship [106].

2.3.3. Analytical and Clinical Verification

The several steps involved in the analytical and clinical verification of a molecular diagnostic assay are described in the Clinical and Laboratory Standards Institute procedures [10, 107] and are comparable to those described for other clinical diagnostics assays (Table **1**) [12, 108]. However, there is no consensus in the literature on the number of clinical samples that should be used for the validation of molecular diagnostic assays [12, 13, 18]. The sample number should be statistically relevant, but for some assays the number of positive samples is limited. The high cost of molecular diagnostic assays is an additional problem for the validation of large sample sets [18]. Definitions of aspects that should be taken into account for technical validation are described by Raymaekers *et al.* [18] and are listed in Table **3**. As for all clinical laboratory testing, results obtained with molecular diagnostic assays should be correlated with results obtained with other methods, preferentially a reference method, and within the clinical context ("clinical validation").

2.3.4. Internal Quality Control

Amplification of an internal control (IC) must be included in every assay to exclude false negative results caused by interference with inhibitors, errors in PCR mixture, equipment malfunction and failure of nucleic acid extraction. Amplification of a host species gene as the IC can be used for cell-rich specimens. On the other hand, for cell-free specimens, a synthetic IC can be added to the specimen before extraction. The IC must be added at a suitable concentration to prevent competition for reagents with the target template. The real-time PCR for IC amplification should be optimized in a way that the target gene amplification is preferential to that of the IC.

Inclusion of a negative control, extracted and analysed simultaneously with the specimens, enables the detection of possible contamination during the extraction or the reaction preparation. A no-template control (NTC) must be used to detect reagent contamination or increased back ground signal. Although the problem of contamination [12, 37] is not part of this chapter, it is important to mention that each laboratory should validate its own decontamination procedures [18].

A statistical follow-up of a positive control (reference material) is necessary to monitor the stability of the assay. The concentration of the control should be near the lower limit of detection but sufficiently high to obtain reproducible results. For quantitative assays, CLSI recommends testing at least two concentrations of reference material: one is for sensitivity control and one for high positive control [10].

The target values are determined by calculating the mean and the corresponding standard deviation on multiple (*e.g.*, 20) measurements on different days [109]. One could consider also the variability attributed to different reagent lots, different instruments, different environmental conditions and different operators [18]. Comparison of oligonucleotide sequences of primers and probe with newly described target sequences in a reference database should be performed on a regular basis because small changes in the target could cause primer and/or probe mismatches, resulting in false negative reactions.

2.3.5. Proficiency Testing

If available, external quality assessment is necessary for each assay that is performed [110]. If an external proficiency program survey is not offered, alternative testing can include blind sample testing, exchange of samples with other laboratories, or medical chart review and is recommended to be conducted twice yearly [111].

Table 3: Definitions describing the technical aspects of a validation

<table>
<tr><td>
<ul>
<li>Precision (inter and intra run)
  <ul>
  <li>➤ <u>Definition:</u> the precision is defined as the level of concordance of the individual test results within a single run (intra-assay precision) and between runs (inter-assay precision) [112].</li>
  <li>➤ <u>Characterized by:</u> standard deviation of the measurements and coefficient of correlation.</li>
  </ul>
</li>
<li>Trueness (accuracy)
  <ul>
  <li>➤ <u>Definition:</u> Trueness is defined as the degree of conformity of a measured or calculated quantity to its actual (true) value and can be estimated by analysis of reference materials or comparisons of results with those obtained by a reference method [112].</li>
  <li>➤ <u>Characterized by:</u> percentage agreement with the reference method/material [12].</li>
  </ul>
</li>
<li>Linearity (measuring range)
  <ul>
  <li>➤ <u>Definition:</u> the linearity is defined as the determination of the linear range of quantification.</li>
  <li>➤ <u>Characterized by:</u> regression coefficient (ideally 1) after linear regression.</li>
  </ul>
</li>
<li>Limit of detection (LOD)/analytical sensitivity
  <ul>
  <li>➤ <u>Definition:</u> the LOD is the lowest concentration or quantity of an analyte where ≥ 95 % of test runs give positive results, following serial dilutions of an international/national reference material (if available), calibrated reference material or sample, tested under routine laboratory conditions (COI = 95 %) [10, 12, 85].</li>
  </ul>
</li>
<li>Limit of quantification (LOQ)
  <ul>
  <li>➤ <u>Definition:</u> the LOQ is the lowest and highest concentration of analyte that can be detected with acceptable precision and accuracy, under routine laboratory conditions. These concentrations establish the measuring range for the assay [10].</li>
  </ul>
</li>
<li>Analytical specificity
  <ul>
  <li>➤ <u>Definition:</u> the analytical specificity is defined as the method's ability to obtain negative results in concordance with negative results obtained by the reference method [10].</li>
  </ul>
</li>
</ul>
</td></tr>
</table>

3. HARMONIZATION OF DIFFERENT METHODOLOGIES

The guiding principle described herein is a critical step in harmonization of different methodologies. Recommendations include the use of sample material and sample processing, use of appropriate standards, reference materials, calibrators and an international scale of measurement [113-115].

Although scientific literature can aid in standardization, it is important to address the issue of integrity of scientific literature data. Recently, Bustin *et al.* proposed the MIQE guidelines, which are a set of rules that describe the minimum information necessary for evaluating quantitative PCR experiments [116].

4. CONCLUSION

Real-time PCR is a technique commonly used in molecular diagnostic laboratories. In-house developed assays must be adequately designed, optimized and validated before being applied to routine diagnostics. This chapter describes practical guidelines for the design, optimization and validation of in-house developed real-time PCR assays. Application of these guidelines will lead to a reliable and robust real-time PCR. Because the guiding principle is independent of environment, equipment, and specific applications, it can be exchanged between laboratories. An ultimate aim should be a universal consensus approach of this nature.

REFERENCES

[1] Saiki RK, Scharf S, Faloona F, *et al.* Enzymatic amplification of beta-globin genomic sequences and restriction site analysis for diagnosis of sickle cell anemia. Science 1985; 230: 1350-4.

[2] Mullis KB, Faloona FA. Specific synthesis of DNA *in vitro via* a polymerase-catalyzed chain reaction. Methods Enzymol 1987; 155: 335-50.

[3]   Higuchi R, Dollinger G, Walsh PS, Griffith R. Simultaneous amplification and detection of specific DNA sequences. Biotechnology (N Y) 1992; 10: 413-7.

[4]   Higuchi R, Fockler C, Dollinger G, Watson R. Kinetic PCR analysis: real-time monitoring of DNA amplification reactions. Biotechnology (N Y) 1993; 11: 1026-30.

[5]   Holland PM, Abramson RD, Watson R, Gelfand DH. Detection of specific polymerase chain reaction product by utilizing the 5'----3' exonuclease activity of Thermus aquaticus DNA polymerase. Proc Natl Acad Sci USA 1991; 88: 7276-80.

[6]   Lee LG, Connell CR, Bloch W. Allelic discrimination by nick-translation PCR with fluorogenic probes. Nucleic Acids Res 1993; 21: 3761-6.

[7]   Livak KJ, Flood SJ, Marmaro J, Giusti W, Deetz K. Oligonucleotides with fluorescent dyes at opposite ends provide a quenched probe system useful for detecting PCR product and nucleic acid hybridization. PCR Methods Appl 1995; 4: 357-62.

[8]   Csako G. Present and future of rapid and/or high-throughput methods for nucleic acid testing. Clin Chim Acta 2006; 363: 6-31.

[9]   Kaltenboeck B, Wang C. Advances in real-time PCR: application to clinical laboratory diagnostics. Adv Clin Chem 2005; 40: 219-59.

[10]   CLSI. Quantitative Molecular Methods for Infectious Diseases; Approved Guideline. NCCLS document MM6-A: NCCLS, 940 West Valley Road, Suite 1400, Wayne, Pennsylvania 19087-1898 USA; 2003.

[11]   Gunson R, Collins T, Carman W. Practical experience of high throughput real time PCR in the routine diagnostic virology setting. J Clin Virol 2006; 35: 355-67.

[12]   Sloan LM. Real-time PCR in clinical microbiology: verification, validation, and contamination control. Clin Microb Newsletter 2007; 29: 87-95.

[13]   Rabenau H, Kessler H, Kortenbusch M, Steinhorst A, Raggam RB, Berger A. Verification and validation of diagnostic laboratory tests in clinical virology. J Clin Virol 2007; 40: 93-8.

[14]   Amos. Technical standards and guidelines for CFTR mutation testing. 2006.

[15]   Sackett DL, Rosenberg WM, Gray JA, Haynes RB, Richardson WS. Evidence based medicine: what it is and what it isn't. BMJ 1996; 312: 71-2.

[16]   Price CP. Evidence-based laboratory medicine: supporting decision-making. Clin Chem 2000; 46: 1041-50.

[17]   Raymaekers M, Smets R, Maes B, Cartuyvels R. Checklist for optimization and validation of real-time PCR assays. J Clin Lab Anal 2009; 23: 145-51.

[18]   Raymaekers M, Bakkus M, Boone E, *et al.* Reflections and proposals to assure quality in molecular diagnostics. Acta Clin Belg 2011; 66: 33-41.

[19]   Ginzinger DG. Gene quantification using real-time quantitative PCR: an emerging technology hits the mainstream. Exp Hematol 2002; 30: 503-12.

[20]   McConlogue L, Brow MA, Innis MA. Structure-independent DNA amplification by PCR using 7-deaza-2'-deoxyguanosine. Nucleic Acids Res 1988; 16: 9869.

[21]   Saiki RK. The design and optimization of the PCR. New York: Stockton Press; 1989.

[22]   Ririe KM, Rasmussen RP, Wittwer CT. Product differentiation by analysis of DNA melting curves during the polymerase chain reaction. Anal Biochem 1997; 245: 154-60.

[23]   Nath K, Sarosy JW, Hahn J, Di Como CJ. Effects of ethidium bromide and SYBR® Green I on different polymerase chain reaction systems. J Biochem Biophys Methods 2000; 42: 15-29.

[24]   Giglio S, Monis PT, Saint CP. Demonstration of preferential binding of SYBR® Green I to specific DNA fragments in real-time multiplex PCR. Nucleic Acids Res 2003; 31: e136.

[25]   Gudnason H, Dufva M, Bang DD, Wolff A. Comparison of multiple DNA dyes for real-time PCR: effects of dye concentration and sequence composition on DNA amplification and melting temperature. Nucleic Acids Res 2007; 35: e127.

[26]   Braasch DA, Corey DR. Locked nucleic acid (LNA): fine-tuning the recognition of DNA and RNA. Chem Biol 2001; 8: 1-7.

[27]   Singh SK, A.A. Koshkin, J. Wengel, P. Nielsen. LNA (locked nucleic acids): synthesis and high-affinity nucleic acid recognition. Chem Comm 1998: 455-6.

[28]   Kutyavin IV, Afonina IA, Mills A, *et al.* 3'-minor groove binder-DNA probes increase sequence specificity at PCR extension temperatures. Nucleic Acids Res 2000; 28: 655-61.

[29]   Tyagi S, Bratu DP, Kramer FR. Multicolor molecular beacons for allele discrimination. Nat Biotechnol 1998; 16: 49-53.

[30]   Tyagi S, Kramer FR. Molecular beacons: probes that fluoresce upon hybridization. Nat Biotechnol 1996; 14: 303-8.

[31]   Caplin BE, Rasmussen RP, Bernard PS, Wittwer CT. LightCycler hybridization probes - the most direct way to monitor PCR amplification and mutation detection. Biochem 1999; 1: 5-8.

[32]   Kusser W. Use of self-quenched, fluorogenic LUX primers for gene expression profiling. Methods Mol Biol 2006; 335: 115-33.

[33]   Svanvik N, Westman G, Wang D, Kubista M. Light-up probes: thiazole orange-conjugated peptide nucleic acid for detection of target nucleic acid in homogeneous solution. Anal Biochem 2000; 281: 26-35.

[34]   Li Q, Luan G, Guo Q, Liang J. A new class of homogeneous nucleic acid probes based on specific displacement hybridization. Nucleic Acids Res 2002; 30: E5.

[35]   Kubista M, Andrade JM, Bengtsson M, *et al.* The real-time polymerase chain reaction. Mol Aspects Med 2006; 27: 95-125.

[36]   Arya M, Shergill IS, Williamson M, Gommersall L, Arya N, Patel HR. Basic principles of real-time quantitative PCR. Expert Rev Mol Diagn 2005; 5: 209-19.

[37]   Aslanzadeh J. Preventing PCR amplification carryover contamination in a clinical laboratory. Ann Clin Lab Sci 2004; 34: 389-96.

[38]   Bustin SA, Nolan T. Pitfalls of quantitative real-time reverse-transcription polymerase chain reaction. J Biomol Tech 2004; 15: 155-66.

[39]   Mackay IM. Real-time PCR in the microbiology laboratory. Clin Microbiol Infect 2004; 10: 190-212.

[40]   Tan BH, Lim EA, Liaw JC, Seah SG, Yap EP. Diagnostic value of real-time capillary thermal cycler in virus detection. Expert Rev Mol Diagn 2004; 4: 219-30.

[41]   Muller M, Erben P, Saglio G, *et al.* Harmonization of BCR-ABL mRNA quantification using a uniform multifunctional control plasmid in 37 international laboratories. Leukemia 2008; 22: 96-102.

[42]   Silvy M, Mancini J, Thirion X, Sigaux F, Gabert J. Evaluation of real-time quantitative PCR machines for the monitoring of fusion gene transcripts using the Europe against cancer protocol. Leukemia 2005; 19: 305-7.

[43]   Gundry CN, Vandersteen JG, Reed GH, Pryor RJ, Chen J, Wittwer CT. Amplicon melting analysis with labeled primers: a closed-tube method for differentiating homozygotes and heterozygotes. Clin Chem 2003; 49: 396-406.

[44]   Wittwer CT, Reed GH, Gundry CN, Vandersteen JG, Pryor RJ. High-resolution genotyping by amplicon melting analysis using LCGreen. Clin Chem 2003; 49: 853-60.

[45]   Vossen RH, Aten E, Roos A, den Dunnen JT. High-resolution melting analysis (HRMA): more than just sequence variant screening. Hum Mutat 2009; 30: 860-6.

[46]   Montgomery JL, Sanford LN, Wittwer CT. High-resolution DNA melting analysis in clinical research and diagnostics. Expert Rev Mol Diagn 2010; 10: 219-40.

[47]   Saiki RK, Gelfand DH, Stoffel S, *et al.* Primer-directed enzymatic amplification of DNA with a thermostable DNA polymerase. Science 1988; 239: 487-91.

[48]   Huang J, DeGraves FJ, Gao D, Feng P, Schlapp T, Kaltenboeck B. Quantitative detection of Chlamydia spp. by fluorescent PCR in the LightCycler. Biotechniques 2001; 30: 150-7.

[49]   Bustin SA. Absolute quantification of mRNA using real-time reverse transcription polymerase chain reaction assays. J Mol Endocrinol 2000; 25: 169-93.

[50]   DeGraves FJ, Gao D, Kaltenboeck B. High-sensitivity quantitative PCR platform. Biotechniques 2003; 34: 106-10, 12-5.

[51]   Longo MC, Berninger MS, Hartley JL. Use of uracil DNA glycosylase to control carry-over contamination in polymerase chain reactions. Gene 1990; 93: 125-8.

[52]   Espy MJ, Smith TF, Persing DH. Dependence of polymerase chain reaction product inactivation protocols on amplicon length and sequence composition. J Clin Microbiol 1993; 31: 2361-5.

[53]   Borst A, Box AT, Fluit AC. False-positive results and contamination in nucleic acid amplification assays: suggestions for a prevent and destroy strategy. Eur J Clin Microbiol Infect Dis 2004; 23: 289-99.

[54]   Kellogg DE, Rybalkin I, Chen S, *et al.* TaqStart Antibody: "hot start" PCR facilitated by a neutralizing monoclonal antibody directed against Taq DNA polymerase. Biotechniques 1994; 16: 1134-7.

[55]   Bartholomeusz A, Tomlinson E, Wright PJ, *et al.* Use of a flavivirus RNA-dependent RNA polymerase assay to investigate the antiviral activity of selected compounds. Antiviral Res 1994; 24: 341-50.

[56]   AppliedBiosystems. Primer Express Software: http://primer-express.software.informer.com.

[57]   Hyndman DL, Mitsuhashi M. PCR primer design. Methods Mol Biol 2003; 226: 81-8.

[58]   Rozen S, Skaletsky HJ. Primer3 on the WWW for general users and for biologist programmers. Totowa, NJ: Humana Press; 2000.

[59]   Rychlik W. OLIGO 7 primer analysis software. Methods Mol Biol 2007; 402: 35-60.

[60]   Suggs SV, Hirose T., Miyake E.H., *et al.* Using purified genes. New York: Academic Press; 1981.

[61]   Bolton ET, Mc Carthy B. A general method for the isolation of RNA complementary to DNA. Proc Natl Acad Sci USA 1962; 48: 1390-7.

[62]   Breslauer KJ, Frank R, Blocker H, Marky LA. Predicting DNA duplex stability from the base sequence. Proc Natl Acad Sci USA 1986; 83: 3746-50.

[63]   Freier SM, Kierzek R, Jaeger JA, *et al.* Improved free-energy parameters for predictions of RNA duplex stability. Proc Natl Acad Sci USA 1986; 83: 9373-7.

[64]   Mitsuhashi M. Technical report: Part 1. Basic requirements for designing optimal oligonucleotide probe sequences. J Clin Lab Anal 1996; 10: 277-84.

[65]   Rychlik W, Rhoads RE. A computer program for choosing optimal oligonucleotides for filter hybridization, sequencing and *in vitro* amplification of DNA. Nucleic Acids Res 1989; 17: 8543-51.

[66]   Wittwer CT, Herrmann MG, Moss AA, Rasmussen RP. Continuous fluorescence monitoring of rapid cycle DNA amplification. Biotechniques 1997; 22: 130-1, 4-8.

[67]   Mitsuhashi M. Technical report: Part 2. Basic requirements for designing optimal PCR primers. J Clin Lab Anal 1996; 10: 285-93.

[68]   Toouli CD, Turner DR, Grist SA, Morley AA. The effect of cycle number and target size on polymerase chain reaction amplification of polymorphic repetitive sequences. Anal Biochem 2000; 280: 324-6.

[69]   Simonsson T, Pecinka P, Kubista M. DNA tetraplex formation in the control region of c-myc. Nucleic Acids Res 1998; 26: 1167-72.

[70]   Zuker M. http://www.bioinfo.rpi.edu/zukerm/cgi-bin/rna-index.cgi.

[71]   Bej AK, Mahbubani MH, Atlas RM. Amplification of nucleic acids by polymerase chain reaction (PCR) and other methods and their applications. Crit Rev Biochem Mol Biol 1991; 26: 301-34.

[72]   Newton C, Graham A. PCR. Oxford, UK 1994.

[73]   NCBI. http://www.ncbi.nlm.gov/blast/Blast.cgi.

[74]   Wheeler DL, Barrett T, Benson DA, *et al.* Database resources of the National Center for Biotechnology Information. Nucleic Acids Res 2007; 35: D5-12.

[75]   Karlin S, Altschul SF. Methods for assessing the statistical significance of molecular sequence features by using general scoring schemes. Proc Natl Acad Sci USA 1990; 87: 2264-8.

[76]   Altschul SF, Gish W, Miller W, Myers EW, Lipman DJ. Basic local alignment search tool. J Mol Biol 1990; 215: 403-10.

[77]   CLMA's Clinical Laboratory Regulation: A guide to CLIA Compliance: Washing G-2 Reports. Washington 1997.

[78]   Bonroy C, Vankeerberghen A, Boel A, De Beenhouwer H. Use of a multiplex real-time PCR to study the incidence of human metapneumovirus and human respiratory syncytial virus infections during two winter seasons in a Belgian paediatric hospital. Clin Microbiol Infect 2007; 13: 504-9.

[79]   Joss AW, Evans R, Mavin S, Chatterton J, Ho-Yen DO. Development of real time PCR to detect Toxoplasma gondii and Borrelia burgdorferi infections in postal samples. J Clin Pathol 2008; 61: 221-4.

[80]   Riffelmann M, Wirsing von Konig CH, Caro V, Guiso N. Nucleic Acid amplification tests for diagnosis of Bordetella infections. J Clin Microbiol 2005; 43: 4925-9.

[81]   Smalling TW, Sefers SE, Li H, Tang YW. Molecular approaches to detecting herpes simplex virus and enteroviruses in the central nervous system. J Clin Microbiol 2002; 40: 2317-22.

[82]   Tang YW, Mitchell PS, Espy MJ, Smith TF, Persing DH. Molecular diagnosis of herpes simplex virus infections in the central nervous system. J Clin Microbiol 1999; 37: 2127-36.

[83]   van der Velden VH, Boeckx N, Gonzalez M, *et al.* Differential stability of control gene and fusion gene transcripts over time may hamper accurate quantification of minimal residual disease--a study within the Europe Against Cancer Program. Leukemia 2004; 18: 884-6.

[84]   CLSI. MM13-A—Collection, Transport, Preparation, and Storage of Specimens for Molecular Methods; Approved Guideline: Clinical and Laboratory Standards Institute, 940 West Valley Road, Suite 1400, Wayne, Pennsylvania 19087-1898, USA; 2005.

[85]   Espy M, Uhl J, Sloan L, *et al.* Real-time PCR in clinical microbiology: applications for routine laboratory testing. Clin Microbiol Rev 2006; 19: 165-256.

[86] Abu Al-Soud W, Radstrom P. Capacity of nine thermostable DNA polymerases To mediate DNA amplification in the presence of PCR-inhibiting samples. Appl Environ Microbiol 1998; 64: 3748-53.

[87] Wilson IG. Inhibition and facilitation of nucleic acid amplification. Appl Environ Microbiol 1997; 63: 3741-51.

[88] Hoorfar J, Wolffs P, Radstrom P. Diagnostic PCR: validation and sample preparation are two sides of the same coin. APMIS 2004; 112: 808-14.

[89] Radstrom P, Knutsson R, Wolffs P, Lovenklev M, Lofstrom C. Pre-PCR processing: strategies to generate PCR-compatible samples. Mol Biotechnol 2004; 26: 133-46.

[90] Livak KJ, Schmittgen TD. Analysis of relative gene expression data using real-time quantitative PCR and the 2(-Delta Delta C(T)) Method. Methods 2001; 25: 402-8.

[91] Pfaffl MW. A new mathematical model for relative quantification in real-time RT-PCR. Nucleic Acids Res 2001; 29: e45.

[92] Pfaffl MW, Horgan GW, Dempfle L. Relative expression software tool (REST) for group-wise comparison and statistical analysis of relative expression results in real-time PCR. Nucleic Acids Res 2002; 30: e36.

[93] Pfaffl MW, Tichopad A, Prgomet C, Neuvians TP. Determination of stable housekeeping genes, differentially regulated target genes and sample integrity: BestKeeper--Excel-based tool using pair-wise correlations. Biotechnol Lett 2004; 26: 509-15.

[94] Gibbs PJ, Cameron C, Tan LC, Sadek SA, Howell WM. House keeping genes and gene expression analysis in transplant recipients: a note of caution. Transpl Immunol 2003; 12: 89-97.

[95] Huggett J, Dheda K, Bustin S, Zumla A. Real-time RT-PCR normalisation; strategies and considerations. Genes Immun 2005; 6: 279-84.

[96] Rasmussen R. Quantification on the LightCycler instrument. Heidelberg: Springer Press 2001.

[97] Wong ML, Medrano JF. Real-time PCR for mRNA quantitation. Biotechniques 2005; 39: 1-11.

[98] Yuan JS, Reed A, Chen F, Stewart CNJ. Statistical analysis of real-time PCR data. Bioinformatics 2006; 7: 85.

[99] Liu W, Saint DA. A new quantitative method of real time reverse transcription polymerase chain reaction assay based on simulation of polymerase chain reaction kinetics. Anal Biochem 2002; 302: 52-9.

[100] Tichopad A, Dzidic A, Pfaffl MW. Improving quantitative real-time RT-PCR reproducibility by boosting primer-linked amplification efficiency Biotechnology Letters 2002; 24: 2053-6.

[101] Tichopad A, Didier A, Pfaffl MW. Inhibition of real-time RT–PCR quantification due to tissue-specific contaminants. Molecular and Cellular Probes 2004; 18: 45-50.

[102] Rutledge RG. Sigmoidal curve-fitting redefines quantitative real-time PCR with the prospective of developing automated high-throughput applications. Nucleic Acids Res 2004; 32: e178.

[103] Tichopad A, Dilger M, Schwarz G, Pfaffl MW. Standardized determination of real-time PCR effciency from a single reaction set-up. Nucleic Aids Research 2003; 31: e122.

[104] Cikos S, Bukovska A, Koppel J. Relative quantification of mRNA: comparison of methods currently used for real-time PCR data analysis. Mol Biol 2007; 8: 113.

[105] Kontanis EJ, Reed FA. Evaluation of Real-Time PCR Amplification Efficiencies to Detect PCR Inhibitors. J Forensic Sci 2006; Vol. 51: 795-804.

[106] Snedecor G, Cochran W. Statistical Methods. seventh ed. Iowa: The Iowa State University Press; 1980.

[107] CLSI. Clinical and Laboratory Standards Institute. Molecular Diagnostic Methods for Infectious Diseases; Approved Guideline—Second Edition. CLSI document MM3-A2: Clinical and Laboratory Standards Institute, 940 West Valley Road, Suite 1400, Wayne, Pennsylvania 19087-1898 USA; 2006.

[108] CLSI. CLSI document CS2.

[109] Westgard. Tietz textbook of clinical chemistry 1999.

[110] CLSI. MM14-A—Proficiency Testing (External Quality Assessment) for Molecular Methods; Approved Guideline: Clinical and Laboratory Standards Institute, 940 West Valley Road, Suite 1400, Wayne, Pennsylvania 19087-1898, USA; 2005.

[111] NCCLS. Assessment of Laboratory Tests When Proficiency Testing is Not Available; Approved Guideline, GP29-A: NCCLS, 940 West Valley Road, Suite 1400, Wayne, Pennsylvania 19087-1898, USA; 2002.

[112] Rabenau HF, Kessler HH, Kortenbusch M, Steinhorst A, Raggam RB, Berger A. Verification and validation of diagnostic laboratory tests in clinical virology. J Clin Virol 2007; 40: 93-8.

[113] Branford S, Cross NC, Hochhaus A, *et al.* Rationale for the recommendations for harmonizing current methodology for detecting BCR-ABL transcripts in patients with chronic myeloid leukaemia. Leukemia 2006 20: 1925-30. .

[114] Hughes T, Deininger M, Hochhaus A, *et al.* Monitoring CML patients responding to treatment with tyrosine kinase inhibitors: review and recommendations for harmonizing current methodology for detecting BCR-ABL transcripts and kinase domain mutations and for expressing results. Blood 2006; 108: 28-37.

[115] Gabert J, Beillard E, van der Velden VH, *et al.* Standardization and quality control studies of 'real-time' quantitative reverse transcriptase polymerase chain reaction of fusion gene transcripts for residual disease detection in leukemia - a Europe Against Cancer program. Leukemia 2003; 17: 2318-57.

[116] Bustin SA, Benes V, Garson JA, *et al.* The MIQE guidelines: minimum information for publication of quantitative real-time PCR experiments. Clin Chem 2009; 55: 611-22.

CHAPTER 3

Antimicrobial Resistance in Bacterial Pathogens: Mechanisms and PCR-Based Detection Technologies

Bashar W. Shaheen[1*], Rajesh Nayak[1] and Dawn M. Boothe[2]

[1]*Division of Microbiology, National Center for Toxicological Research, US Food and Drug Administration, Jefferson, AR, USA and* [2] *Department of Anatomy, Physiology and Pharmacology, College of Veterinary Medicine, Auburn University, Auburn, AL, USA*

Abstract: In the last decade, antimicrobial resistance has been widespread in several bacterial species. This increase in resistance could be associated with an increase in the use of different antimicrobials to treat infections caused by pathogenic bacteria. While resistance to antimicrobials is often attributed to known mechanisms, other mechanisms are still under investigation for many bacterial species. Detection of antimicrobial resistance often involves conventional agar, broth or disk diffusion assays. However, these methods can be cumbersome and time consuming compared to molecular methods. Consequently, several polymerase chain reaction (PCR) techniques have been developed to expedite the detection of antimicrobial resistance in bacterial pathogens. PCR-based technologies are rapid, sensitive and specific for detecting antimicrobial resistance. Application of such technologies in diagnostic laboratories can provide insight into emerging mechanisms of antimicrobial resistance in veterinary pathogens. In this chapter, we describe molecular mechanisms of drug resistance in microbial pathogens and the potential advantages and disadvantages of PCR-based methods.

Keywords: Antimicrobial resistance; polymerase chain reaction; mechanisms of resistance; susceptibility testing; molecular methods; bacterial pathogens; antimicrobials; DNA sequencing; detection.

1. INTRODUCTION

In veterinary medicine, antimicrobial agents have been used historically to treat infections or for prophylaxis. In the United States, at least 17 classes of antimicrobials have been approved for use in food animals [1]. A large proportion of these antimicrobials have been used in feed additives for metaphylaxis. In companion animals, several classes of antimicrobial drugs, such as cephalosporins (first and third generation), other β-lactams (amoxicillin with or without clavulanic acid), tetracyclines and fluoroquinolones have been approved for use [2]. Furthermore, the Animal Medicinal Drug Use act legalized extra-label drug use by veterinarians (http://www.fda.gov/AnimalVeterinary/GuidanceComplianceEnforcement/ActsRulesRegulations/ucm085377. htm). As a result, most antimicrobials approved for use in humans also have been used, to varying degrees, for treatment of infections in either food or companion animals. This use has been associated with an increased trend toward antimicrobial resistance in animal pathogens [3-6], and may impose limitations on the therapeutic options to treat infections caused by resistant microorganisms. Therefore, good eradication programs, vaccination and hygiene practices are important to minimize diseases caused by resistant pathogens. Furthermore, judicious use of antimicrobial agents can improve the overall health conditions of diseased animals and may reduce the emergence of multi-drug resistant pathogens. Most importantly, using appropriate dosing regimens to treat infection, another challenge for veterinarians, could minimize the potential risk of antimicrobial resistance.

The emergence of antimicrobial resistance in veterinary medicine is a major public health concern. This can be attributed to resistant bacteria, which causes disease in humans, and can be transmitted *via* contaminated food [7]. In one study, molecular subtyping of quinolone-resistant *Campylobacter jejuni* strains provided evidence of an association between the strains isolated from chicken products and those isolated from

Address correspondence to Bashar W. Shaheen: U.S. Food and Drug Administration, National Center for Toxicological Research, Division of Microbiology, Jefferson, AR 72079-9502, USA; Tel: 870-543-7599, Fax: 870-543-7307; E-mail: bashar.shaheen@fda.hhs.gov

Chengming Wang, Bernhard Kaltenboeck and Mark D. Freeman (Eds)

domestically acquired infections in Minnesota residents [8]. Studies have also suggested direct transmission of resistant organisms from food-producing and companion animals to humans [2, 9]. The risk may be escalated through close physical contact between household pets and humans [2]. Antimicrobial resistance genes in bacteria from companion animals share identical genetic profiles with resistant strains found in humans [2]. Pets may acquire and transfer multidrug-resistant pathogens to humans, particularly methicillin-resistant *S. aureus* [10-12] and *Escherichia coli* [13-15]. Resistance to selected drug classes (*e.g.*, fluoroquinolones, extended-spectrum cephalosporins, and glycopeptides) is of particular concern because of their importance in human medicine.

Judicious use of antimicrobials is difficult to define and even more difficult to apply. Among the more important tenets is de-escalation of antimicrobial use, particularly those which are unnecessary. However, if a decision is made to use an antimicrobial, then drugs should be used with a "hit hard and get out quick" approach [16]. The most suitable drug is the one that is very narrow in its spectrum, ideally targeting the infecting organism and minimally impacting other organisms. In animals previously exposed to antimicrobials, drugs should be selected based, accordingly, on their susceptibility profiles [16]. Dosing regimens should be designed based on integration of pharmacokinetic and pharmacodynamic data to ensure that sufficient quantities of a drug reach the site of infection to eliminate even those colony forming units with high minimum inhibitory concentrations (MICs). Sub-therapeutic drug concentrations could select for isolates with increased levels of resistance [17].

An abbreviated, rather than a prolonged, course of antimicrobial therapy is emerging as an approach to limit the opportunity for developing resistance by the bacteria [18, 19]. Susceptibility testing and appropriate design of dosing regimens are increasingly being promoted in companion animal practices as means of decreasing resistance [20]. Bacteria are characterized by either inherent or acquired resistance to antimicrobials. The latter can develop through chromosomal mutations and can be reduced through optimization of the dosing regimens. Acquired resistance, which may also be conferred by transmissible mobile genetic elements, might be minimized by avoiding direct contact and subsequent selective pressure. These are some examples of approaches to prudent antimicrobial use by veterinary practitioners if current antimicrobials are to be preserved for future use.

2. MOLECULAR MECHANISMS OF ANTIMICROBIAL RESISTANCE

2.1. Resistance to Fluoroquinolones

2.1.1. Resistance Through Chromosomal Mutations

Fluoroquinolones inhibit DNA synthesis by stabilizing breaks in bacterial DNA made by DNA gyrase or topoisomerase IV [21]. The mutations in DNA gyrase (two subunits: GyrA and GyrB) and topoisomerase IV (two subunits: ParC and ParE) are the most common mechanisms that confer resistance to quinolones [22-24]. While topoisomerase II is the primary target of quinolone resistance in Gram-negative isolates and the secondary target in Gram-positive isolates, topoisomerase IV is the primary target in Gram-positive bacteria and a secondary target in Gram-negative bacteria [25-28]. Most mutations are located in a region called the quinolone resistance-determining region (QRDR) of *gyrA*. Amino acid changes in the QRDR result in altered protein structure of the quinolone binding site near the interface of the enzyme, reducing the quinolone affinity for the modified enzyme-DNA complex [23]. QRDR regions in *E. coli* are located between amino acids Ala67 and Gln106 in *gyrA* [29] and amino acids Asp426 and Lys447 in *gyrB* [30]. Mutations that confer resistance to quinolones, as a result of altered binding at the active site of *gyrA,* are most commonly found in clinical, veterinary and laboratory strains of *E. coli* and occur at codons 83 (predominantly) and/or 87 of the *gyrA* gene [21, 22, 31-35]. In *E. coli*, mutation at codon 83 generally involves a substitution of serine with leucine, conferring high-level resistance (MIC value=0.39 - 3.13 μg/ml) to fluoroquinolones and their progenitors, including nalidixic acid [32, 36, 37]. However, additional mutations in the *gyrA* and/or the topoisomerase IV genes are essential to confer higher-level resistance (MIC value=6.25 to 100 μg/ml) to fluoroquinolones [22, 36, 37]. In *Salmonella,* the mutations occur mostly at codons Ser83 and Asp87, although other mutations have also been identified at codons Ala67, Asp72, Gly81 and Asp82 [21]. In addition, other mutations have been characterized outside the QRDR in salmonellae, including a mutation at codon Ala119 to Ser, Glu or Val [21] and at codons Ala131, Glu139 and Asp144 [21, 38]. One study indicated that amino acid change Asp87Gly was the most common mutation found in the veterinary salmonellae isolates [39]. However, another study indicated that most of the veterinary *S.* Newport isolates had mutations at codon Ser83 [40]. Similarly, Ser83Phe mutation was commonly observed within *S.* Typhi and *S.* Paratyphi A [41] and as a secondary target mutation in farm animals isolates [21, 39].

2.1.2. Altered and Protected Drug Target Site

The plasmid mediated quinolone resistance (PMQR) gene *qnrA* offers protection against quinolone inhibition by binding directly to both subunits and the holoenzyme of either gyrase or topoisomerase IV, thereby inhibiting the gyrase-DNA interaction [42]. This process can destabilize the lethal gyrase-DNA-quinolone cleavage complex. Another PMQR enzyme is an aminoglycoside acetyltransferase, encoded by *aac(6')-Ib*, which confers resistance to tobramycin, amikacin and kanamycin. However, the ciprofloxacin resistance variant *aac(6')-Ib-cr* has also been shown to *N*-acetylate ciprofloxacin and norfloxacin at the amino nitrogen of the piperazinyl substituent [43]. This gene is apparently widespread in many geographical areas in the US [44]. Almost 51% of isolates in Shanghai harbored the cr-variant, whereas 11% carried the wild type aminoglycoside acetyltransferase [45]. Among 313 *Enterobacteriaceae* surveyed in the United States, the wild variant (*aac(6')-Ib*) was present in 50.5% of isolates whereas 28% carried the cr variant [44].

In 1994, the *qnrA* gene was first identified in *Klebsiella pneumoniae* from a patient's urine sample in Alabama [46]. Since then, many epidemiological surveys have been conducted to detect its presence in Gram-negative bacteria, including *E. coli, Enterobacter cloacae, Providencia stuartii, Citrobacter freundii, Citrobacter koseri, Shigella flexneri* and non-Typhi *Salmonella enterica* [44]. Many variations in the protein have been identified that differ in the amino acid sequence composition from the original *qnrA* gene [47, 48]. This includes variants *qnrA, qnrB, qnrS, qnrC* and *qnrD* [44, 49]. It is believed that these genes have originated from *Shewanella* species, *Vibrio vulnificus, Vibrio parahaemolyticus* and *Photobacterium profundum* isolated from water samples [50, 51]. The quinolone MIC levels for these organisms were four-fold to eight-fold higher than for isolates lacking these genes.

The *qepA*-mediated efflux system is another PMQR that was more recently discovered in *E. coli* in Belgium and Japan [52, 53]. It mediates resistance to hydrophilic quinolones (*i.e.*, norfloxacin, ciprofloxacin or enrofloxacin) by acting as a proton antiporter efflux pump system that increases the excretion of fluoroquinolones [53]. The *mfpA* gene, which has been identified in *Mycobacterium tuberculosis*, interacts with DNA gyrase and interferes with fluoroquinolones' inhibitory action [43]. Although a low level of resistance is conferred by the PMQR mechanism, it substantially enhances the number of resistant mutants that can be selected from the population. A sensitive strain of *E. coli* carrying *qnrA*, but devoid of mutation(s) in gyrase or topoisomerases, developed chromosomal mutations and subsequent high-level resistance after five days of norfloxacin treatment [54]. More recently, *qnrB* or *qnrS* were detected in 14 (6%) of the veterinary poultry and swine clinical isolates of *E. coli* [55]. In addition, the *aac (6')-Ib-cr* variant allele gene was first detected in two *qnr* isolates from pigs and ducks [55]. PMQR determinants were present in 35 (34.7%) isolates, with *qnr, aac (6')-Ib-cr*, and *qepA* detected alone or in different combination in ceftiofur-resistant *Enterobacteriaceae* isolates from companion and food-producing animals [56].

2.1.3. Efflux Pump System

Five different families of efflux pump proteins have been identified in Gram-positive and -negative bacteria. These include the resistance nodulation division family (RND), the major facilitator superfamily (MFS), the staphylococcal multi-resistance (SMR) plasmids, the multidrug and toxic compound extrusion families (MATE) and the ATP binding cassette (ABC) [57]. Some efflux pumps have narrow ranges of substrate profiles; they confer high levels of resistance to tetracycline [58] chloramphenicol and macrolides [59], and are mediated by mobile genetic elements. In contrast, multidrug efflux pumps can act on, and transport, multiple structurally unrelated drugs, thus contributing to the emergence of multidrug resistant phenotypes [57]. The tripartite RND multidrug efflux pumps are important in *E. coli* and other Gram-negative bacteria (*e.g., E. coli acrB*/AcrB, *P. aeruginosa mexB*/MexB, *Campylobacter jejuni cmeB*/CmeB and *Neisseria gonorrhoeae mtrD*/MtrD). The RND pumps are proton antiporters that power drug efflux by exchanging H$^+$ ions across the membrane gradient [60]. The AcrAB-TolC system is an example of an RND system in *E. coli* [57]. It has a wide range of substrates besides fluoroquinolones, including chloramphenicol, lipophilic β-lactams, tetracycline, rifampin, novobiocin, fusidic acid, nalidixic acid, ethidium bromide, acriflavine, bile salts, short-chain fatty acids, SDS, Triton X-100 and triclosan [61-65]. Individually, over-expression of AcrAB-TolC system or mutations in topoisomerase genes is unlikely to increase clinical levels of resistance to selected drugs, such as ciprofloxacin, chloramphenicol, tetracycline or cotrimoxazole [57, 66]. However, a combination of mutation coupled with over-expression of an efflux pump has been shown to increase fluoroquinolone resistance [67-69]. Furthermore, one study showed the absence of mutations in regulatory genes (*i.e.*, *marRAB* or *soxRS*) in *acrB*-over expressing fluoroquinolone-resistant clinical isolates of *E. coli* from veterinary sources [70]. However, mutation at amino acid 45 of AcrR, that changed arginine with cysteine, increased the expression of AcrAB and the sensitivity to ciprofloxacin among clinical and veterinary isolates of *E. coli* [71]. Similarly, veterinary isolates of *S.* Typhimurium elicited an increase in AcrB over-expression which is associated with MICs' increase to fluoroquinolone above the recommended clinical and laboratory standards institute (CLSI) breakpoint [72-74].

2.2. Resistance to β-Lactam Antibiotics

2.2.1. Enzymatic Drug Modification

The active site of β-lactams is the β-lactam ring. The ring substitutes for a penicillin-binding protein (PBP) substrate, thus interfering with cross linking of the peptidoglycan, which is necessary for cell wall synthesis [75]. The most common mechanism whereby bacteria acquire resistance to β-lactams is the production of β-lactamases, which hydrolyze the β-lactam ring, rendering the drug incapable of binding to the PBPs active site [76, 77]. Several β-lactamases have been characterized [78]. Resistance to β-lactams in Gram-negative bacteria is mediated either by chromosomally located β-lactamase encoding genes or those acquired through mobile genetic elements, such as plasmids or transposons [76, 77]. The *ampC* gene, which is found in chromosomes of *Enterobacteriaceae* and *P. aeruginosa*, also confers resistance to β-lactams through plasmid-mediated *ampC* genes in *E. coli*, *Klebsiella* species, and *Salmonella* species [78]. Mutations in *ampC* promoter, which have been recognized among isolates from animals [79, 80], increase production of *Amp*C and confer resistance to penicillins, monobactams and cephalosporins [77, 79, 81]. One study showed that 8 out of 18 cefazolin-resistant *E. coli* strains from chickens had mutations in the *Amp*C promoter region [82]. Other β-lactamase genes were found in the chromosomes of multiple drug resistant *Salmonella* species isolated from food animals, which encoded for enzymes PSE-1 (CARB-2) and OXA-30 (OXA-1) [83, 85]. These β-lactamase genes are active against penicillins and first/second generation cephalosporins [77]. These resistance genes were also identified on mobile elements and in a region called the *Salmonella* genomic island (SGI1), associated with genes conferring resistance to aminoglycosides, chloramphenicol/florfenicol, sulfonamides, and tetracyclines [86, 87]. Plasmid-borne β-lactamase genes are widely distributed among animal-derived *E. coli* and *Salmonella* species [88]. The CMY β-lactamases enzymes, which were first identified in *Klebsiella pneumoniae* (cephamycinases), were considered extended spectrum β-lactamases (ESBLs) (*i.e.*, β-lactamases that hydrolyze oxyimino-cephalosporins and can be inhibited by clavulanate) [88, 89]. The CMY-2 enzyme is widespread in extended-spectrum cephalosporin-resistant *E. coli* and *Salmonella* species of animal origin, and their association with integrons/transposons may facilitate their dissemination [90, 91]. Over 200 ESBLs have been documented in Gram-negative isolates from humans [88]; these include TEM, SHV, OXA and CTX-M β-lactamases [76, 78, 92]. The CTX-M family of ESBLs is prevalent among animal isolates and confers high levels of resistance to aminopenicillins, carboxypenicillins, ureidopenicillins, narrow-spectrum first/second generation cephalosporins, third generation cephalosporins and variable levels of resistance to fourth generation cephalosporins [93]. CTX-M β-lactamase-encoding genes are harbored in plasmids with other ESBL-genes, such as *bla*$_{TEM}$ genes that confer resistance to aminoglycosides, chloramphenicol, sulfonamides, trimethoprim, and tetracyclines [93, 94]. The CTX-M β-lactamases are emerging among *E. coli* and *Salmonella* isolates from food and companion animals [88]. Other plasmid-encoded β-lactamases, including TEM-type and SHV-type enzymes, have been identified in β-lactam-resistant *E. coli* and *Salmonella* species from animals [95-97].

2.2.2. Altered and Protected Drug Target Site

Shortly after the approval of methicillin in 1959, methicillin-resistance was first identified in *Staphylococcus aureus* [98]. The penicillin-binding proteins (PBPs) play an important role in the resistance to β-lactams in *S. aureus*, Enterococcus species and *Streptococcus pneumoniae* [99]. The resistance is caused by acquisition of exogenous genes that alter PBP, resulting in resistance to inhibition by β-lactams. The altered PBP2a, which is encoded by *mecA,* is mediated by mobile elements called the staphylococcal cassette chromosome (SCC*mec*) [100]. Usually PBP2 in *S. aureus* encodes for both transpeptidase (PBP2a) and transglycosylase (PBP2) activity [101]. The transpeptidase function of PBP2a enables the bacteria to become resistant to β-lactams. Resistance to β-lactams in *S. aureus* was also due to alteration of penicillin binding affinity to PBP1, PBP2 and elevation of PBP4 [99]. Methicillin-resistant *S. aureus* (MRSA) causes life-threatening infections, and is widely distributed in the U.S. [102, 103]. The increase in the level of resistance in human isolates of *Enterococcus faecium* is associated with an increase in expression of PBP5, and mutations of PBP5 further decrease its affinity for β-lactams [104, 105]. Thus, the high level of resistance to β-lactams is due to combined mechanisms of both overexpression and reduced affinity [104, 105]. In clinical isolates of *Streptococcus pneumoniae*, the modified PBP1a, PBP2b, PBP2x and PBP2a have been found to less effectively bind radio-labeled β-lactams in the resistant clinical isolates than the susceptible ones [99, 106]. The over-expressed PBPs in *Enterococcus faecium* and *Enterococcus faecalis* decrease the affinity for penicillin and confer resistance to ampicillin. In addition, resistance to β-lactams can be mediated through the expression of multidrug efflux pumps [107, 108]. Unlike other mechanisms of resistance, the level of resistance conferred by efflux systems is low [109]. This mechanism has been recognized in *E. coli* and *Salmonella* species from food animals [110, 111].

2.3. Resistance to Tetracyclines

2.3.1. Active Efflux Pump

The Tet proteins, which belong to the major facilitator superfamily (MFS), include more than 300 different proteins [112]. Their role is to efflux tetracycline out of the bacterial cell thereby protecting the ribosomes. The efflux pump proteins have been characterized into six groups based on amino acid sequence identity [112]. The efflux system is a proton antiporter that exchanges a proton for a tetracycline molecule across the concentration gradient of the membrane [113]. In Gram-negative bacteria, the efflux genes are associated with large conjugative plasmids belonging to different plasmid incompatibility groups [114, 115]. Usually these plasmids harbor other antimicrobial resistance genes, heavy metal resistance genes and virulence factors [112]. Thus, these mobile genetic elements can contribute to the emergence of multi-drug-resistant phenotypes among bacteria [116, 117].

2.3.2. Altered and Protected Drug Target Site

The mechanism of resistance to tetracyclines involves production of cytoplasmic ribosome protection proteins (RPPs) that protect the ribosomes from the action of tetracycline, doxycycline and minocycline [112, 118]. The two RPPs Tet(O) and Tet(M) are widely distributed and have been isolated from *Campylobacter jejuni* and *Streptococcus* species [112]. The RPPs have great similarity to the elongation factors EF-Tu and EF-G [119, 120]. These RPPs bind to the ribosome and cause conformational changes that dislodge tetracyclines from binding to their binding site without change in protein synthesis [112]. The ribosomal conformational change is energy dependent and requires GTP hydrolysis [118]. Tet(M), which has high affinity to bind to ribosomes compared to the elongation factor EF-G, must dissociate first from the ribosome to allow EF-G to bind [121]. The resistance determinant Tet(P) from *Clostridium* species has two overlapping genes: one encodes for efflux protein and the other for tetracycline RPPs [122]. Enzymatic drug modification has been described as another mechanism of imparting resistance to tetracycline. A good example is the *tet*(X) gene found in transposons of the anaerobic *Bacteroides* [123]. The flavin-dependent monooxygenase, Tet(X), which modifies tigecycline to form 11a-hydroxytigecycline, inhibits protein translation through the formation of a weaker complex than tigecycline with magnesium (which is important for binding tetracycline) [124].

In veterinary medicine, tetracycline has been used to treat infection in companion animals [125] and as a growth promoter [126, 127]. The United States Food and Drug Administration approved the use of this drug as a feed additive in 1951 (chlortetracycline) and 1953 (oxytetracycline) [112]. Several studies have indicated that the use of tetracycline for veterinary therapeutics, and animal growth promotion results in the selection of resistant bacterial pathogens including zoonotic pathogens such as *Salmonella* and *Campylobacter* spp., and commensals such as *Escherichia coli* and enterococci [128-131]. The risk associated with the use of tetracycline as a growth promoter is the selection of resistance due to application of this drug at subtherapeutic levels for long term usage instead of short-term usage at concentrations high enough to treat infectious disease in animals [112, 132]. In addition, studies have suggested the use of tetracycline in veterinary medicine can induce the transfer of antimicrobial resistance gene(s) of *Salmonella* species from animals to humans [131, 133, 134].

2.4. Resistance to Sulfonamides and Trimethoprim

2.4.1. Resistance Through Chromosomal Mutations

Sulfonamides, coupled with diaminopyridines (*e.g.*, trimethoprim), are competitive inhibitors of enzymes responsible for synthesis of folate, a substrate that is necessary for DNA synthesis. Sulfonamides inhibit dihydropteroate synthase (DHPS) and diaminopyridine inhibits dihydrofolate reductase (DHFR). The combination of these two drugs is bactericidal. In human isolates of *C. jejuni* and *Haemophilus influenzae,* mutations in the gene encoding for dihydropteroate synthase decrease its affinity for sulfonamide binding, whereas mutations in the gene encoding dihydrofolate reductase result in over expression of the enzyme, producing an altered secondary structure of the enzyme and consequently reduced affinity for trimethoprim [135]. A single amino acid substitution in DHPS enzyme can result in resistance to sulfonamides in *E. coli, S. aureus, Staphylococcus haemolyticus, C. jejuni* and *Helicobacter pylori* [136], whereas a single amino acid substitution in DHFR causes resistance to trimethoprim in *S. aureus* [137] and *S. pneumonia* [138]. Additionally, mutations at the promoter region of chromosomal *dhfr* genes have been described in resistant isolates of *H. influenza* [139].

2.4.2. Plasmid Mediated Sulfonamides and Trimethoprim Resistance

The DHPS enzymes are encoded by *sul*I and *sul*II genes in Gram-negative bacteria [135, 136]. The *sul*I gene was found on conjugative plasmids as part of class 1 integrons of transposon Tn*21*, whereas the *sul*II gene was found on conjugative or non-conjugative plasmids [136, 140]. Different *dhfr* genes, which confer resistance to trimethoprim, were found as part of gene cassettes in Gram-negative bacteria [141]. The *dfr*1 gene was found in cassettes in both class 1 and class 2 integrons in Gram-negative bacteria [135]. A plasmid mediated trimethoprim resistance gene, *dfr*9, which was originally detected in swine *E. coli* isolates [142], has been found in trimethoprim-resistant veterinary and human isolates of *E. coli* [143]. Their prevalence has been attributed to high quantities of prescriptions for trimethoprim in the swine industry [135]. The *dfr* genes (*dfrA1, dhfrA17, and dfrA12*), mediating trimethoprim resistance, were found in class 1 integron among canine and feline clinical isolates of *E. coli* [144].

2.5. Resistance to Aminoglycosides

2.5.1. Enzymatic Drug Modification

Aminoglycoside-modifying enzymes that confer resistance in human isolates are mostly encoded in mobile elements [145]. Enzymatic drug modification by *N*-acetyltransferases (AACs), *O*-phosphotransferases (APHs), and *O*-nucleotidyltransferases (ANTs) cause structural changes in the aminoglycosides, modifying binding to target ribosomal sites [146, 147]. The majority of AAC enzymes have been described exclusively in Gram-negative bacteria [148]. Soon after the approval of apramycin for use in veterinary medicine, isolates of *E. coli* and *Salmonella* were found to be resistant to apramycin and gentamicin due to the enzyme AAC(3)-IV [149, 150]. Several variants of ANTs and APHs have been described that differ in their substrate specificity and confer different resistant phenotypes [145, 146, 148, 151-153]. Genes encoding for production of these enzymes (*aac, ant and aph*) are generally present on mobile genetic elements, such as plasmids, transposons, or integrons [145-148, 154]. Their acquisition and dissemination is usually associated with other antimicrobial resistance genes, such as genes conferring resistance to carbapenems, and sulfonamides [155], or chloramphenicol, trimethoprim and streptothricin [144], resulting in multi-drug resistant phenotypes [116].

Other mechanisms, such as a change in the charge of cell wall lipopolysaccharides have been shown to prevent the positively charged aminoglycosides from crossing the outer membrane [141]. Since the drug requires an electron transport system to cross the cytoplasmic membrane, anaerobes and facultative anaerobes are inherently resistant to aminoglycosides [141]. In addition, specific efflux pumps, such as AcrD in *E. coli* [156] and MexXY in *Pseudomonas* [157] have been shown to decrease the concentration of aminoglycosides inside the cell.

2.6. Resistance to Chloramphenicol and Florphenicol

2.6.1. Enzymatic Drug Inactivation

In veterinary medicine, chloramphenicol has been approved to treat infections in pets and non-food producing animals in the European Union and North America [158]. Enzymatic drug inactivation is the most common mechanism of resistance to chloramphenicol in Gram-positive and Gram-negative bacteria [159, 160]. The mechanism is mediated through chloramphenicol acetyltransferases (CAT), which transfer acetyl groups to the C1 and C3 positions of the chloramphenicol molecule; the modified molecules are unable to inhibit bacterial protein biosynthesis [159, 160]. Two distinct types of CAT enzymes have been recognized including CAT-A and CAT-B enzymes [159]. The *catI, catII* and *catIII* genes are widespread among Gram-negative bacteria [161-162]. They have been identified in mobile genetic elements such as transposons (Tn9) and plasmids [163-165]. The *catII* gene has been identified mainly in *Haemophilus* species [162, 166], whereas the *catIII* gene was detected in *Enterobacteriaceae* and *Pasteurella* [141]. The *cat*(A-E) genes, which encode for acetylating enzymes, confer resistance to streptogramin [159]. The *catB* gene was detected on the chromosome of *Agrobacterium tumefaciens*, *P.aeruginosa* or *Vibrio cholera*, *E. coli* transposon Tn*2424* and *Morganella morganii* transposon Tn*840* [141]. The *catB* gene was identified in the gene cassettes of class 1 integron in *E. coli* isolated from dogs and cats in the US [144].

2.6.2. Decreased Accumulation

Active chloramphenicol transporters, mediated by *cml*A and *cml*B genes, have been identified in *Pseudomonas aeruginosa* and *Rhodococcus fascians* [60]. The resistance mechanism is mediated by mobile genetic element transposon Tn*1696* [154]. Multidrug transporter systems, such as MexAB/OprM and MexCD/OprJ in *P. aeruginosa* and AcrAB/TolC in *E. coli* and *Salmonella* export chloramphenicol [167-169]. Reduced permeability is another mechanism that confers resistance to different drug classes. The bacterial outer membrane provides an additional barrier that prevents the entry of hydrophilic and hydrophobic antimicrobials into bacteria. This mechanism plays an important role in *P. aeruginosa* and to a lesser extent in *E. coli* [170]. The resistance, mediated through the outer membrane proteins (OmpF in *E. coli* [171] and OmpK36 porin in *Klebsiella pneumonia* [172] and OprD in *P. aeruginosa* [65]), mutations (mar-phenotype) or decreased porin expression, confers low level resistance to tetracycline, chloramphenicol and norfloxacin [173].

3. STANDARD METHODS FOR SUSCEPTIBILITY TESTING

Several *in vitro* methods, such as disk diffusion, broth dilution, and agar dilution have been used to measure antimicrobial susceptibility in bacteria. The Clinical and Laboratory Standards Institute (CLSI) [174] recommends performance standards, application protocols, interpretive standards and limitations of the standard testing methods. The performance standards' publications are updated regularly for a wide range of bacterial isolates [174]. There are several challenges facing any changes or modifications of these procedures to determine new interpretive criteria. These include (i) that the method requires validation and studies on animals to reflect *in vivo* efficacy and (ii) the difficulty in developing new quality controls and the interpretation of the expected clinical outcome in animal models [174]. The limitations encountered when applying *in vitro* testing data to *in vivo* models can be avoided through studies that investigate the effects and mechanisms of action of antimicrobial agents on animals.

The disk diffusion and broth dilution assays have been applied for fast growing bacteria, such as *Staphylococcus* species, *Pasteurella* species or *Enterobacteriaceae* and fastidious bacteria, such as

Actinobacillus pleuropneumoniae, *Haemophilus* species and *Campylobacter jejuni* [174]. For fastidious microorganisms, additional precautions are required. For example, the media for disk diffusion testing for *Haemophilus somnus* and *A. pleuropneumoniae* should be supplemented with additional ingredients (hemoglobin and nutritional supplements) and some modifications are warranted in the procedures [174]. Similarly, the Mueller-Hinton agar media for *Streptococcus* species should be supplemented with 5% defibrinated sheep blood [174].

3.1. Advantages of Standard Susceptibility Testing Methods

Susceptibility testing supports both the selection of drugs and the design of dosing regimens for treatment of infectious diseases. Additionally, the susceptibility profiles can provide updated epidemiological data, especially for newly approved antimicrobials.

3.2. Disadvantages of Standard Susceptibility Testing Methods

Current culture and susceptibility testing is intended to support therapeutic use of antimicrobials. However, relatively few antimicrobial drugs have established interpretive criteria in animals. Rather, testing of many drugs approved for use in animals is based on human interpretive criteria, even when establishing standards for treating animals, due to the lack of information specific for the target animal. Susceptibility testing requires isolation of the infecting microbe. Accordingly, it is not possible to directly test clinical samples, such as body fluid or urine samples. Results are markedly dependent on standardized procedures, with variability influenced by method. Satisfactory training and educational programs are essential for staff and clinicians to understand the quantitative results obtained from the test.

The standard susceptibility testing involves several steps, which are tedious and time consuming. This imposes a time limitation on clinicians, especially with critical care patients, to prescribe or design the dosing regimen based on the results obtained from the standard susceptibility testing. In addition, quality control isolates are needed to perform the standard susceptibility testing. Misleading conclusions can occur when combined antimicrobials are tested against certain bacterial species (*e.g.*, first and second generation cephalosporins and aminoglycosides against *Salmonella* species) or when the emergence of resistance occurs as a result of lengthy therapy (*e.g.*, *Pseudomonas aeruginosa* with all drugs and staphylococci with fluoroquinolones) [174].

4. MOLECULAR METHODS FOR EVALUATING THE ANTIMICROBIAL RESISTANCE

4.1. Advantages of Molecular Susceptibility Testing Methods

Genetic methods of susceptibility testing offer several advantages to standard susceptibility testing. First, molecular methods can be applied directly to clinical samples (sterile body fluid or urine samples) and do not require organism isolation and purification. Second, genetic methods detect changes in the genomic content of the organism, while the standard susceptibility testing monitors the changes in phenotypic expression of the organism. Because many factors influence phenotype, the specificity of genotypic description facilitates organism identity. Further, some factors can affect the results obtained from the conventional susceptibility techniques such as, growth conditions, handling of the isolates and potential cross contamination. Genetic methods are quantitative and measure the relative intensities of amplicon signals, depending on the initial concentration of the target DNA. In real time PCR, the amplification process can be monitored during the reaction process in real time. Third, the genotypic methods have been proven to be specific in determining the changes that may occur in the target genes. Therefore, specific genotypic information can provide precise information regarding the mechanisms contributing to emerging resistance. Lastly, a major advantage of genetic methods is the potential to detect a low copy number of the target gene; a particular advantage when detecting the gene in clinical samples with low numbers. These methods allow less time lapse between sample collection and result acquisition, compared with standard susceptibility testing, which usually requires 3 to 4 days after receiving the clinical samples to obtaining the final MIC values.

4.2. Disadvantages of Molecular Susceptibility Testing Methods

There are some disadvantages with molecular techniques for susceptibility testing. First, multiple genes are required for detecting different antimicrobial agents associated with various mechanisms. Therefore, several optimization protocols are required to determine the antimicrobial resistance pattern. Second, different mechanisms may contribute to the expression of a particular resistant phenotype; hence, standard susceptibility testing needs to be conducted along with the molecular methods. Other mechanisms besides mutations could play a critical role in the emergence of a resistant phenotype [23, 57]. Standard susceptibility testing may be required to provide a complete picture of the phenotypic expression, especially when some resistance determinants are not clearly defined. Third, nucleic acid cross contamination may occur from external sources. However, this problem has been solved by adding chemicals or enzymes to the PCR reaction (*i.e.*, adding uracil-N-glycosylase to prevent product carryover contamination) [175]. Balancing the concentration of reagents and the temperature of the PCR cycles is important to get reliable results. Sequencing and multi-sequence alignment data are necessary before designing the primers and the probes. In addition, genetic methods require expensive instrumentation, such as a LightCycler® or Smart-Cycler® to perform these assays, which may not be available in routine diagnostic laboratories.

4.3. PCR-Based Methods Used for Detection of Antimicrobial Resistance

The basic concept of PCR is to amplify a known sequence or short segment of the genomic DNA with designed primers [176, 177]. The procedure involves several thermal cycles and a DNA polymerase to synthesize a single strand of DNA from the 3' -OH end of each primer. The number of copies of the target sequence increases exponentially and corresponds to the cycle number. It doubles with each cycle until reaching a plateau at which no more primer-template accumulates and then the increase in the target DNA becomes linear. The final PCR product is confirmed by different methods, such as agarose gel electrophoresis, restriction fragment length polymorphism (RFLP), probe hybridization assays or DNA sequencing. Several methods have been adapted to detect the gene mutations associated with antimicrobial resistance. These PCR based methods include RFLP, single-strand conformation polymorphism (SSCP), cleavase fragment length polymorphism (CFLP), mismatch amplification mutation assay-PCR (MAMA-PCR) and real time PCR assay [178-180]. PCR-based methods have been widely used in veterinary medicine to detect multiple antimicrobial resistance genes (Table 1).

4.3.1. Conventional PCR and DNA Sequencing

Conventional PCR has been used historically to detect antimicrobial resistance genes [181-183]. Amplification of the target sequence depends on careful design of primers specific to the gene of interest. The optimization of primers can be time consuming. The amplicon lengths for PCR are variable with 50-60% GC content. Usually the amplification of short fragments is more efficient than that of a longer fragment. To insure specific amplification, one should avoid stretches of three or more GC nucleotides at the 3' end of primers. The melting temperature (T_m), which is defined as the temperature at which a given DNA fragment exists as 50% each double- and single-stranded DNA, is dependent on the length and GC content and the degree of homology between the two DNA strands [175]. In addition to adjusting the annealing temperature, the primers should be between 20 to 30 bp, with melting temperatures (T_m) ranging between 60°C and 72°C. A higher T_m range of the primers can provide a specific annealing temperature, which insures specific amplification. In general, the technique offers a powerful tool for distinguishing individual alleles in a genome, and thus the ability to diagnose a disease that is defined at the sequence level. It also allows DNA to be amplified from very small tissue samples [184]. The method has been used in veterinary medicine to detect antimicrobial resistance genes in bacterial pathogens (Table 1).

Table 1: PCR-based methods used in veterinary medicine and their characteristics

PCR Based Methods	Potential Advantages and Disadvantages	Antimicrobial Drug Classes	Resistant Organism and Host (Reference)
Conventional PCR and sequencing	Advantages: simple, straight forward and requires inexpensive instrumentation. The method has high discriminatory power Disadvantages: post PCR sequencing is costly and involves using expensive instrumentation. Cross contamination can occur	Fluoroquinolones β-lactams Tetracycline Aminoglycosides, sulfonamide, chloramphenicol, trimethoprim	*Enterobacter* spp. in dogs and cats [265]; *Salmonella* in chickens [203] *S. aureus* in dogs, cats [266], cattle, pigs and cats [267]; *Salmonella* and *E. coli* in chickens [268] *E. coli* in calve [269]

		and aminoglycosides	
			E. coli in cattle [270]
Multiplex PCR	Advantages: ability to simultaneously detect multiple genes of interest in a single reaction, inexpensive and simple instrumentation	β-lactams	*S. aureus* in pig [271]
		β-lactams, and tetracycline	*E. coli* in steers [272]
		Aminoglycosides, sulfonamides, tetracyclines and chloramphenicols	Fecal bacteria in giant pandas [262]
	Disadvantages: needs optimization of the PCR conditions		*S. Enteritidis* and *S. Typhimurium* in cattle [273]
		β-lactams, chloramphenicol and tetracycline	
RFLP-PCR	Advantages: simple and can be used in many applications.	Fluoroquinolones	*Mycoplasma bovis* in cattle [274]; *Helicobacter* in dogs and cats [275]; *Klebsiella pneumoniae* in dogs, cats, cattles, birds, monkey, elephants, seal and guinea pigs [276]; *E. coli* in dogs and cats [277]
		β-lactams	
	Disadvantages: examines only a small portion of the genomic DNA and less discriminatory power	Class 1 integrons	*Salmonella Typhimurium* in horses [278]
			E. coli in cattle, dogs and cats [144, 211]
SSCP-PCR	Advantages: less time consuming, cheap, simple to perform and can differentiate the mutant from the wild allele based on the mobility	Fluoroquinolones	*C. jejuni* in chickens [279]
	Disadvantages: moderate discriminatory power since no information regarding the location of the mutation is provided.		
MAMA-PCR	Advantages: Simple, fast and requires inexpensive instrumentation.	Fluoroquinolones	*Salmonella* in pigs and chickens [231]; *Campylobacter* in pigs, broilers and humans [232]
	Disadvantages: the presence of novel, different mutations or those outside the region of interest cannot be detected.		
SYBR Green	Advantages: rapid, inexpensive and quantitative.	Bacitracin and streptogramin	*E. coli* and *Enterococcus* species in broilers chickens broilers and chickens [280]
	Disadvantages: primer-dimers cause nonspecific amplification.		
FRET probe	Advantages: rapid, specific and sensitive method, with the ability to detect small copy number of the gene	Fluoroquinolones	*E. coli* in dogs and cats [34]; Salmonella Typhimurium in turkeys, cattle, dogs, cats, pigs, rabbits, pheasants, partridges and wild birds [248]
	Disadvantages: designing the primer and the probe is required, and optimization of the PCR conditions are important to get reliable results		

Multiplex PCR involves simultaneous detection of multiple antimicrobial resistance genes in a single reaction tube [7, 182, 185-188]. Furthermore, housekeeping genes can be included in the same PCR reaction, which provide information about bacterial identification [182]. Although the size of the amplicon can provide an initial assessment of the gene prevalence, final confirmation will depend on the DNA sequence information of the purified amplicon. Primers for multiplex PCR are typically designed to have a close annealing temperature. The multiplex PCR has been used to detect methicillin resistance in staphylococci, including the coagulase-negative staphylococci related to the *mecA* gene [182, 189, 190]. A novel quadriplex PCR assay has been developed to simultaneously discriminate *S. aureus* from other *Staphylococcus* species, including the less virulent coagulase-negative organisms, and determine their antimicrobial resistance (*i.e.*, methicillin and mupirocin resistance) [188]. Quadriplex PCR has both high specificity and sensitivity, increasing its potential clinical importance as a rapid detection method. More recently, a multiplex PCR-ligation detection reaction assay has been developed for the simultaneous detection of multi-drug resistance and toxin genes from *S. aureus*, *Enterococcus faecalis* and *Enterococcus faecium* [191]. This assay exhibits high specificity and sensitivity (>94%) for rapid detection of multi-drug resistance and virulence genes. In veterinary medicine, multiplex PCR was used for characterization of multiple drug resistant *S. Typhimurium*, which are resistant to chloramphenicol, streptomycin, sulfisoxazole and tetracycline, and identification of β-lactamase gene distribution among *Salmonella* isolated from fecal samples of cattle [192]. Additional examples are shown in Table **1**.

Conventional PCR is cheaper than other molecular methods. Heat block instruments used for conventional PCR have the economic advantage of accommodating a 96-well sample format, which is more economic than some real-time PCR instrumentation. However, accuracy can be a major problem due to uneven temperature distribution across the heat block [188, 193]. Furthermore, heat block instrumentation requires more time to reach the target temperature, and as a result more time to complete one reaction. However, these deficiencies have been addressed by introducing new versions of the heat block machines and the use of heated air of real-time PCR instruments. For conventional PCR, a post amplification process is required, using gel electrophoresis and staining, to confirm the identity as the target DNA sequence. Cross contamination and product carryover can occur with this method. The sensitivity of conventional PCR may be affected by replication errors, which may occur in preliminary cycles. Hence, sequencing of the amplicon is essential for final confirmation of amplicon identity. Sequencing can be expensive and requires sophisticated instrumentation.

4.3.2. PCR-Based Methods Used for Detection of Mutations Associated with Antimicrobial Resistance

PCR-Restriction Fragment Length Polymorphism (RFLP)

This method involves the ability of a restriction endonuclease, which is sequence dependent, to enzymatically digest the DNA and separate different DNA fragments that can be visualized in an agarose gel [180]. The restriction endonucleases should have specific recognition sites of either four or six bases in a palindromic sequence. In the presence of DNA mutation at the restriction site, polymorphism can be observed by missing or gaining DNA bands making it possible for the mutants to be differentiated using an agarose gel. Therefore, the ability of RFLP to differentiate different alleles is limited to the mutations that occur within the recognition site that the restriction endonuclease would digest [180]. The method has been proven to be discriminatory for identifying mutations associated with antimicrobial resistance. PCR-RFLP has been used to identify isoniazid resistant catalase-peroxidase (*katG*)-associated mutations in *M. tuberculosis* [194-201]. Using the restriction endonuclease *Msp*I, the mutations were located in the hot spots of codons 315 (change serine to threonine) and 463 (change arginine to leucine) [197, 198, 200]. This method was rapid, cost-effective, and provided high sensitivity and specificity (*i.e.*, 80% and 100%, respectively) against standard susceptibility testing for rapid identification of isoniazid-resistant *M. tuberculosis* [201]. RFLP has been used as a rapid screening test for detection of gyrase genes associated with fluoroquinolone resistance in *Streptococcus pneumoniae*, *Salmonella* species, *Campylobacter coli*, *S. aureus* and *E. coli* [36, 202-207]. The RFLP-PCR is also used to detect Leu-83 and Gly-87 mutations in *gyrA* among quinolone-resistant clinical isolates of *E. coli* from humans [208]. In veterinary medicine, PCR-RFLP of the *groEL* gene was successfully used for molecular identification of coagulase-negative *Staphylococcus* species responsible for bovine mastitis [209]. In veterinary and human medicine, RFLP of *flaA* PCR fragments (*flaA*-RFLP) has been used to study *Campylobacter* epidemiology [210]. RFLP has been applied to characterized class 1 integrons in *E. coli* strains isolated from bovine mastitis in Mongolia [211] and in *E. coli* from food-producing animals in The Netherlands [9]. More recently, our laboratory has studied RFLP-PCR for molecular characterization of class 1 and 2 integrons in canine and feline clinical isolates of *E. coli* in the US [144] (Table **1**).

PCR- Single-Strand Conformation Polymorphism (SSCP)

This technique has been used to detect gene mutation associated with resistance to β-lactams in *Enterobacteriaceae* [212], and isoniazid [213, 214], rifampin [215, 216], ethambutol [217] and fluoroquinolones in *M. tuberculosis* [218]. The basic principle of the technique is that the PCR product is denatured to single stranded DNA by alkali treatment and then electrophoresed using neutral polyacrylamide gels [180]. The band differences can be used to differentiate between wild type and mutant strains [219]. These differences are due to the mutation-altered secondary structure of the PCR amplicon (*i.e.*, the altered ssDNA folded structure of the mutant) [220]. Unlike RFLP, the technique is useful to differentiate the whole PCR amplicon independent of the specific restriction site [180]. Smaller amplicons (310 to 320 bp) usually produce more sensitive results [213, 221]. Use of radioactive labeling for detection of the PCR product is one disadvantage of this technique. Modifications have included fluorescein labeling [222] and silver-staining methods [213]. Nested PCR-linked SSCP analysis and DNA sequencing have been used to detect *M. tuberculosis* and determine rifampin (RIF) susceptibility directly from clinical sputum samples [223]. This method combined with sequence analyses of the RNA polymerase gene (*rpoB*) has been used for simultaneous identification and discrimination between RIF-resistant *Mycobacterium tuberculosis* and non-tuberculous mycobacteria [109]. A new multi-PCR-SSCP assay was developed to detect isoniazid (INH) and RIF resistance in *M. tuberculosis* in a single reaction [224]. The assay was specific (100% and 92%, for INH and RIF, respectively), rapid and inexpensive for detecting resistant isolates. More recently, this method has been modified to identify the mutations associated with pyrazinamide resistance in *M. tuberculosis* directly from sputum samples [225]. The method showed a high percentage of agreement (*i.e.*, ≥ 90%) with other methods, and is cost effective and rapid (less than 24 h) compared with other assays. In veterinary medicine, the SSCP-PCR was used to analyze *gyrA* gene in *Salmonella* serotypes isolated from farm animals [39]. SSCP methodology has also been applied to detect nucleotide changes of *parE* gene in ciprofloxacin-resistant clinical human isolates of *E. coli* from Argentina and Spain, and veterinary isolates from the United Kingdom [22].

PCR- Cleavase Fragment Length Polymorphism (CFLP)

Cleavase I restriction endonuclease introduces cleavage at the junction of the single-stranded DNA and duplexed areas (*i.e.*, structures formed by base-paired DNA at elevated temperatures [194]), resulting in a distinct restriction pattern [178]. The process first involves the formation of unique hairpin structures, which occur during the annealing temperature as a result of self-base-pairing of single strands of DNA [226]. Therefore, a series of stem-loop structures are formed and connected by non-base-paired single-stranded regions. Cleavase I recognizes the structures and cleaves at the junction of the single-stranded DNA and duplexed areas [226]. The pattern difference can be recognized by gain or loss of a DNA band, differences in the movement on the gel and intensity of the band. The discriminatory power of this method is moderate and dependent on the DNA sequence of the locus examined. SSCP and CFLP methodologies have detected the mutations of exon VIII in the human iduronate 2-sulfatase (IDS) gene associated with X-linked lysosomal storage disease, Hunter syndrome [227]. The study showed that SSCP was able to detect 100% of the mutations in the IDS gene, whereas CFLP could detect only 50% of the mutations. However, another study indicated that CFLP was more sensitive in identifying mutations in short sequences of *rpoB* mutants of *M. tuberculosis* [228]. Thus, the discriminatory power of this method is variable and potentially dependent on the DNA sequence of a gene. Furthermore, the method has shown to have poor reproducibility [228, 229], be more difficult to perform than PCR-RFLP and interpret, and be more time consuming [226]. These disadvantages probably limit the use of CFLP in veterinary and human medicine as a method for detection of antimicrobial resistance in bacterial pathogens. The method has been used to detect mutations in the *katG* and *rpoB* genes of *M. tuberculosis* [194, 228].

Mismatch Amplification Mutation Assay (MAMA)-PCR

MAMA-PCR was first developed as a rapid and specific method for detection of *gyrA* mutations associated with ciprofloxacin-resistant *C. jejuni* [179]. In MAMA-PCR, the forward and reverse mutation primers are used to amplify a 265-bp PCR product, an indication of the Thr-86-to-Ile (ACA→ATA) mutation in the *C. jejuni gyrA* gene [179]. A negative result (no mutation) was indicated by the absence of an amplicon in the gel. In addition, another conserved reverse primer was used with the forward primer to generate a 368 bp fragment, which was a positive control for the *C. jejuni gyrA* gene [179]. The method detects *gyrA* and mutations associated with ciprofloxacin-resistant clinical isolates of *E. coli* and *Neisseria gonorrhoeae* [192, 230]. In food animals, the method has been used to detect *parC* mutations associated with ciprofloxacin resistance in *Salmonella* [231] and *Campylobacter* [232].

Real Time PCR Format

dsDNA-Binding Dyes (SYBR Green): Real time PCR has been developed to quantify the PCR process as it occurs using either an intercalating dsDNA dye (*i.e.*, SYBR Green) or a DNA hybridization probe (*i.e.*, fluorescence resonance energy transfer) (see chapter 2 for complete description of the technology) [233-

235]. The binding affinity of SYBR Green I to dsDNA was more sensitive than ethidium bromide [236]. This dye is used in real time PCR because the fluorescence was greatly enhanced by binding to dsDNA [175]. During the various stages of PCR, different intensities of fluorescence signals can be detected, depending on the amount of dsDNA in the sample. In the elongation phase of PCR, the PCR primers are extended and more SYBR Green I dye is bound to the bacterial DNA, and at the end of the elongation phase, the entire DNA becomes double-stranded and a maximum amount of dye is bound. The fluorescence is recorded at the end of the elongation phase, and increasing amounts of PCR product can be monitored from cycle to cycle. The amplification process can be monitored during the reaction in real time, unlike conventional PCR. The quantitative advantage of this method allows information of the PCR process to be obtained by plotting the intensity of the fluorescence signal versus the cycle number [175]. Furthermore, it is one of the more sensitive methods which can detect as low as a single target copy [236]. The SYBR Green method is inexpensive compared with other real-time PCR formats in which probes are labeled with fluorescent dyes. This method has been applied successfully for rapid detection of methicillin-resistant *S. aureus* (MRSA) from clinical samples, *tetR* of Tn10 in *E. coli* and macrolide resistance in *S. pneumoniae* and *S. pyogenes* [237-239]. One of the disadvantages is that SYBR Green I dye can bind to double-stranded by-products, such as primer dimers, causing non-specific amplification [175].

Hybridization Probe (Fluorescence Resonance Energy Transfer format-FRET): The hybridization probe format utilizes maximum sequence-specificity for detection of the PCR product. The basic principle of the technique is the use of two standard DNA primers [175, 235]. For FRET PCR, two fluorescently labeled hybridization probes bind to target DNA between the flanking primers. The sequence of the two probes can hybridize to the target sequences on the amplified DNA fragment in a head-to-tail arrangement. The donor dye (*e.g.*, fluorescein) is excited by the blue light source and emits green fluorescent light at a slightly longer wavelength. When the two dyes are in close proximity, the energy emitted excites the acceptor dye attached to the second hybridization probe, which then emits fluorescent light at a different wavelength. The intensity of the emitted light is measured at a certain channel of the LightCycler's optical unit at the end of each annealing step, when it will be at its maximum. This energy transfer, referred to as fluorescence resonance energy transfer (FRET), is highly dependent on the close proximity between the two dye molecules (1 to 5 nucleotides). The amount of fluorescence is directly proportional to the amount of target DNA generated during the PCR process. After annealing, the temperature is raised and the hybridization probe is displaced during the elongation step. At the end of this step, the PCR product is double-stranded and the displaced hybridization probes are too far apart to allow FRET to occur.

The hybridization probes bind perfectly to the matching target DNA. The more tightly bound the probe, the higher the T_m required for separation. Thus, binding of the probe to the sequence devoid of mutation(s) yields a higher melting temperature than binding to mutated DNA, which contains destabilizing mismatches. A rapid PCR-based FRET method has been developed to detect ciprofloxacin-resistance in Yersinia pestis *strains* that carry a mutation in codon 81 or codon 83 of gyrA [240] and quinolone-resistant *Neisseria gonorrhoeae* strains in urine samples [241]. Because the stability of hybridization varies with the phenotype, melting temperatures consequently vary. Discriminating among the different phenotypes that carry mutation(s) is indicated by melting curve analysis (MCA). Therefore, amplicon identification and sequencing generally are not necessary, precluding the need for post-PCR processing. This minimizes the possibility of contamination, as amplification and genotyping will be performed in the same sealed tube. However, some disadvantages have been recognized. This method requires more optimization than other methods. Balancing the concentrations of reagents and temperatures of PCR cycles are important to acquire reliable results. Furthermore, sequencing and multi-sequence alignment of data are required before designing the primers and the probes.

We applied real-time PCR assay to detect fluoroquinolone resistance levels in clinical canine and feline isolates of *Escherichia coli* [34]. In this experiment, we differentiated *E. coli gyrA* mutants-enrofloxacin resistant (ENR^r) from wild-type susceptible isolates. Our results indicated that the amplicons of the resistant-type *gyrA* isolates formed a less-stable hybrid with the Bodipy 630/650 probe than did the susceptible ones. The ENR^r isolates exhibited T_m of 61°C, whereas the susceptible isolates exhibited a T_m of 71°C [34]. The results also indicated that the FRET-PCR assay can directly detect single nucleotide polymorphisms (SNP) in *E. coli gyrA* from urine samples from companion animals. Because fluoroquinolone resistance may result from mechanisms other than mutations in *gyrA* or *parC* genes (*e.g.*, mutations in the regulatory genes and their effect on the constitutive overexpression of efflux pumps AcrAB-TolC system and porins or PMQR), standard susceptibility testing was required, especially for those isolates that had no mutations in *gyrA* or *parC*.

In addition, the FRET-PCR format detected mutations in genes associated with quinolone resistance in *C. coli*, *mecA* in *S. aureus*, vancomycin-resistant enterococci (VRE) in clinical specimens, and isoniazid (*katG*, *inhA* and *ahpC*) in *M. tuberculosis* [242-245]. Other FRET PCR probes, such as molecular beacons, have been used to detect fluoroquinolone resistance associated with *grlA* mutations in *S. aureus,* and methicillin resistance associated with *mecA* mutations in *S. aureus* [246, 247]. This technique has been used to detect *gyrA* mutations associated with fluoroquinolone resistance in multi-drug resistant *S. Typhimurium* DT104 from animal origin [248].

Hydrolysis Probe (TaqMan): The Taqman probe is another form of real time PCR in which short oligonucleotides are labeled with a 5'-terminal reporter dye and a 3'-terminal fluorescence quencher [249]. The quencher dye absorbs the fluorescence from the reporter and therefore, no emission can be observed. During the extension phase of PCR , the 5'-exonuclease activity of *Taq* polymerase hydrolyses the probe which hybridizes to the target sequence [249]. Therefore, the fluorescent emissions are released from the separated reporter dye and quencher so that they are detectable after excitation [234]. The method has been used to detect methicillin-resistant *Staphylococcus aureus* from clinical patients [250], and drug-resistant mutations in *rpoB*, *katG*, and *embB* genes in *M. tuberculosis* in clinical patients [251]. Furthermore, a duplex TaqMan real-time PCR assay has been used for rapid detection of ampicillin-resistant *Enterococcus faecium* in clinical patients [252]. In veterinary medicine, TaqMan method was used for identifying ciprofloxacin-resistant *C. jejuni* strains carrying mutation at codon 86 of *gyrA* [253]. There is limited information available on using this technique in veterinary medicine to detect mutation associated with antimicrobial resistance in bacterial pathogens. This was attributed to the fact that melting curve analysis was not possible with this format [175] and the need to design multiple probes that correspond to each mutation can add further drawbacks to this technique [251-253].

Non PCR-Based Method: Microarray Technology

In microarray, the DNA probe is immobilized onto a solid surface-support and hybridized with a fluorescently labeled target (the unknown DNA sequence) [254]. The fluorescence signal intensity increases with increased hybridization between the target and the probe, which can be detected using a fluorescence scanner. Different types of microarray technologies have been developed, including printed double-stranded DNA and oligonucleotide arrays, *in situ*-synthesized arrays, high-density bead arrays, electronic microarrays and suspension bead arrays [254]. Microarray has been applied for epidemiological investigation and characterization, including determination of resistance determinants and virulence factors of methicillin-resistant *S. aureus* (MRSA) [255]. In addition, this technology has been used for the detection of gene mutations associated with antimicrobial resistance in many bacterial species [256-259]. High-density DNA oligonucleotide arrays have been used to detect mutations involved in rifampin resistance in *M. tuberculosis* [260]. In veterinary medicine, the method has been applied to detect antimicrobial resistance genes in methicillin-resistant *Staphylococcus aureus* ST398 from diseased swine [261], *Salmonella* isolates from turkey flocks [262] and *E. coli* isolates from chickens [263]. Some microarray formats are economical, rapid and require little instrumentation [264]. However, ongoing mutations under natural evolution or selective pressure may impose some challenges in a clinical setting. Unlike real-time PCR, further processing steps, such as hybridization and washing, and the potential risk of cross contamination may add further pitfalls for using this technology in the diagnostic laboratory.

5. CONCLUSIONS

This chapter provides insight on the mechanisms and methods for assessing antimicrobial resistance in bacterial pathogens. Some of the genetic tests have yet to be used for routine diagnostic testing. As some techniques provide specificity related to the mechanism of resistance, detection of some antimicrobial resistance determinants will need further evaluation. The standard susceptibility testing can provide a comprehensive picture of the phenotypic expression for certain organisms toward a given drug. However, from a clinical perspective, utilizing rapid, specific, sensitive and reproducible PCR-based tests can be beneficial for both the clinicians and the overall health of the patients or the animals. The molecular techniques provide an alternative to the conventional susceptibility testing. The PCR technology has moved from conventional to real-time, allowing quantification of resistance genes. In addition, molecular methods can aid in early diagnosis of an illness directly from clinical specimens and help physicians and veterinarians accurately prescribe the appropriate antimicrobials for treatment. As the PCR-based technologies continue to evolve, it is likely that rapid genetic methods will be adopted by diagnostic laboratories and hospitals for routine testing of antimicrobial resistance in bacterial pathogens.

ACKNOWLEDGEMENTS

We thank Drs. Carl Cerniglia, John Sutherland and Steven Foley for their useful comments and suggestions. This study was supported in part by appointment (Bashar W. Shaheen) to the Postgraduate Research Program at the National Center for Toxicological Research, administered by the Oak Ridge Institute for Science and Education through an interagency agreement between the US Department of Energy and the US Food and Drug Administration. The views presented in this chapter do not necessarily reflect those of the US Food and Drug Administration.

REFERENCES

[1]   Anderson AD, Nelson JM, Rossiter S, Angulo FJ. Public health consequences of use of antimicrobial agents in food animals in the United States. Microb Drug Resist 2003; 9: 373-9.

[2]   Guardabassi L, Schwarz S, Lloyd DH. Pet animals as reservoirs of antimicrobial-resistant bacteria. J Antimicrob Chemother 2004; 54: 321-32.

[3]   Kruse H, Hofshagen M, Thoresen SI, *et al*. The antimicrobial susceptibility of *Staphylococcus* species isolated from canine dermatitis. Vet Res Commun 1996; 20: 205-14.

[4]   Marano NN, Rossiter S, Stamey K, *et al*. The National Antimicrobial Resistance Monitoring System (NARMS) for enteric bacteria, 1996-1999: surveillance for action. J Am Vet Med Assoc 2000; 217: 1829-30.

[5]   Pellerin JL, Bourdeau P, Sebbag H, Person JM. Epidemiosurveillance of antimicrobial compound resistance of *Staphylococcus intermedium* clinical isolates from canine pyodermas. Comp Immunol Microbiol Infect Dis 1998; 21: 115-33.

[6]   Zhao S, Blickenstaff K, Glenn A, *et al*. Beta-Lactam Resistance in *Salmonella* Isolated from Retail Meats in the United States: National Antimicrobial Resistance Monitoring System (NARMS): 2002-2006. Appl Environ Microbiol 2009.

[7]   Webber M, Piddock LJ. Quinolone resistance in *Escherichia coli*. Vet Res 2001; 32: 275-84.

[8]   Smith KE, Besser JM, Hedberg CW, *et al*. Quinolone-resistant *Campylobacter jejuni* infections in Minnesota, 1992-1998. Investigation Team. N Engl J Med 1999; 340: 1525-32.

[9]   Box AT, Mevius DJ, Schellen P, Verhoef J, Fluit AC. Integrons in *Escherichia coli* from food-producing animals in The Netherlands. Microb Drug Resist 2005; 11: 53-7.

[10]   Loeffler A, Boag AK, Sung J, *et al*. Prevalence of methicillin-resistant *Staphylococcus aureus* among staff and pets in a small animal referral hospital in the UK. J Antimicrob Chemother 2005; 56: 692-7.

[11]   O'Mahony R, Abbott Y, Leonard FC, *et al*. Methicillin-resistant *Staphylococcus aureus* (MRSA) isolated from animals and veterinary personnel in Ireland. Vet Microbiol 2005; 109: 285-96.

[12]   Tomlin J, Pead MJ, Lloyd DH, *et al*. Methicillin-resistant *Staphylococcus aureus* infections in 11 dogs. Vet Rec 1999; 144: 60-4.

[13]   Johnson JR, Owens K, Gajewski A, Clabots C. *Escherichia coli* colonization patterns among human household members and pets, with attention to acute urinary tract infection. J Infect Dis 2008; 197: 218-24.

[14]   Johnson JR, Clabots C. Sharing of virulent *Escherichia coli* clones among household members of a woman with acute cystitis. Clin Infect Dis 2006; 43: e101-8.

[15]   Johnson JR, Brown JJ, Carlino UB, Russo TA. Colonization with and acquisition of uropathogenic *Escherichia coli* as revealed by polymerase chain reaction-based detection. J Infect Dis 1998; 177: 1120-4.

[16] Boothe DM. Culture and susceptibility testing in the critical patient: Perks and Pitfalls. . Journal of Veterinary Emergency and Critical Care 2010; (in Press).

[17] Lesar TS, Rotschafer JC, Strand LM, Solem LD, Zaske DE. Gentamicin dosing errors with four commonly used nomograms. JAMA 1982; 248: 1190-3.

[18] Tam VH, Louie A, Fritsche TR, *et al.* Impact of drug-exposure intensity and duration of therapy on the emergence of *Staphylococcus aureus* resistance to a quinolone antimicrobial. J Infect Dis 2007; 195: 1818-27.

[19] Marra AR, de Almeida SM, Correa L, *et al.* The effect of limiting antimicrobial therapy duration on antimicrobial resistance in the critical care setting. Am J Infect Control 2009; 37: 204-9.

[20] Boothe DM. Principles of antimicrobial therapy. Vet Clin North Am Small Anim Pract 2006; 36: 1003-47, vi.

[21] Hopkins KL, Davies RH, Threlfall EJ. Mechanisms of quinolone resistance in *Escherichia coli* and *Salmonella*: recent developments. Int J Antimicrob Agents 2005; 25: 358-73.

[22] Everett MJ, Jin YF, Ricci V, Piddock LJ. Contributions of individual mechanisms to fluoroquinolone resistance in 36 *Escherichia coli* strains isolated from humans and animals. Antimicrob Agents Chemother 1996; 40: 2380-6.

[23] Piddock LJ. Mechanisms of fluoroquinolone resistance: an update 1994-1998. Drugs 1999; 58 Suppl 2: 11-8.

[24] Vila J, Ruiz J, Goni P, De Anta MT. Detection of mutations in *parC* in quinolone-resistant clinical isolates of *Escherichia coli*. Antimicrob Agents Chemother 1996; 40: 491-3.

[25] Heisig P. Genetic evidence for a role of *parC* mutations in development of high-level fluoroquinolone resistance in *Escherichia coli*. Antimicrob Agents Chemother 1996; 40: 879-85.

[26] Khodursky AB, Zechiedrich EL, Cozzarelli NR. Topoisomerase IV is a target of quinolones in *Escherichia coli*. Proc Natl Acad Sci USA 1995; 92: 11801-5.

[27] Kumagai Y, Kato JI, Hoshino K, *et al.* Quinolone-resistant mutants of *Escherichia coli* DNA topoisomerase IV *parC* gene. Antimicrob Agents Chemother 1996; 40: 710-14.

[28] Ferrero L, Cameron B, Manse B, *et al.* Cloning and primary structure of *Staphylococcus aureus* DNA topoisomerase IV: a primary target of fluoroquinolones. Mol Microbiol 1994; 13: 641-53.

[29] Yoshida H, Bogaki M, Nakamura M, Nakamura S. Quinolone resistance-determining region in the DNA gyrase *gyrA* gene of *Escherichia coli*. Antimicrob Agents Chemother 1990; 34: 1271-2.

[30] Yoshida H, Bogaki M, Nakamura M, Yamanaka LM, Nakamura S. Quinolone resistance-determining region in the DNA gyrase *gyrB* gene of *Escherichia coli*. Antimicrob Agents Chemother 1991; 35: 1647-50.

[31] Vila J, Ruiz J, Marco F, *et al.* Association between double mutation in *gyrA* gene of ciprofloxacin-resistant clinical isolates of *Escherichia coli* and MICs. Antimicrob Agents Chemother 1994; 38: 2477-9.

[32] White DG, Piddock LJ, Maurer JJ, *et al.* Characterization of fluoroquinolone resistance among veterinary isolates of avian *Escherichia coli*. Antimicrob Agents Chemother 2000; 44: 2897-9.

[33] Giraud E, Leroy-Setrin S, Flaujac G, *et al.* Characterization of high-level fluoroquinolone resistance in *Escherichia coli* O78:K80 isolated from turkeys. J Antimicrob Chemother 2001; 47: 341-3.

[34] Shaheen BW, Wang C, Johnson CM, Kaltenboeck B, Boothe DM. Detection of fluoroquinolone resistance level in clinical canine and feline *Escherichia coli* pathogens using rapid real-time PCR assay. Vet Microbiol 2009; 139: 379-85.

[35] Willmott CJ, Maxwell A. A single point mutation in the DNA gyrase A protein greatly reduces binding of fluoroquinolones to the gyrase-DNA complex. Antimicrob Agents Chemother 1993; 37: 126-7.

[36] Ozeki S, Deguchi T, Yasuda M, *et al.* Development of a rapid assay for detecting *gyrA* mutations in *Escherichia coli* and determination of incidence of gyrA mutations in clinical strains isolated from patients with complicated urinary tract infections. J Clin Microbiol 1997; 35: 2315-9.

[37] Weigel LM, Steward CD, Tenover FC. *gyrA* mutations associated with fluoroquinolone resistance in eight species of *Enterobacteriaceae*. Antimicrob Agents Chemother 1998; 42: 2661-7.

[38] Eaves DJ, Liebana E, Woodward MJ, Piddock LJ. Detection of *gyrA* mutations in quinolone-resistant *Salmonella enterica* by denaturing high-performance liquid chromatography. J Clin Microbiol 2002; 40: 4121-5.

[39] Piddock LJ, Ricci V, McLaren I, Griggs DJ. Role of mutation in the *gyrA* and *parC* genes of nalidixic-acid-resistant salmonella serotypes isolated from animals in the United Kingdom. J Antimicrob Chemother 1998; 41: 635-41.

[40] Griggs DJ, Gensberg K, Piddock LJ. Mutations in gyrA gene of quinolone-resistant *Salmonella* serotypes isolated from humans and animals. Antimicrob Agents Chemother 1996; 40: 1009-13.

[41] Walker RA, Skinner JA, Ward LR, Threlfall EJ. LightCycler *gyrA* mutation assay (GAMA) identifies heterogeneity in GyrA in *Salmonella* enterica serotypes Typhi and Paratyphi A with decreased susceptibility to ciprofloxacin. Int J Antimicrob Agents 2003; 22: 622-5.

[42]   Tran JH, Jacoby GA, Hooper DC. Interaction of the plasmid-encoded quinolone resistance protein Qnr with *Escherichia coli* DNA gyrase. Antimicrob Agents Chemother 2005; 49: 118-25.

[43]   Strahilevitz J, Jacoby GA, Hooper DC, Robicsek A. Plasmid-mediated quinolone resistance: a multifaceted threat. Clin Microbiol Rev 2009; 22: 664-89.

[44]   Park CH, Robicsek A, Jacoby GA, Sahm D, Hooper DC. Prevalence in the United States of *aac(6')-Ib-cr* encoding a ciprofloxacin-modifying enzyme. Antimicrob Agents Chemother 2006; 50: 3953-5.

[45]   Robicsek A, Strahilevitz J, Jacoby GA, *et al.* Fluoroquinolone-modifying enzyme: a new adaptation of a common aminoglycoside acetyltransferase. Nat Med 2006; 12: 83-8.

[46]   Martinez-Martinez L, Pascual A, Jacoby GA. Quinolone resistance from a transferable plasmid. Lancet 1998; 351: 797-9.

[47]   Hata M, Suzuki M, Matsumoto M, *et al.* Cloning of a novel gene for quinolone resistance from a transferable plasmid in *Shigella flexneri* 2b. Antimicrob Agents Chemother 2005; 49: 801-3.

[48]   Nordmann P, Poirel L. Emergence of plasmid-mediated resistance to quinolones in *Enterobacteriaceae*. J Antimicrob Chemother 2005; 56: 463-9.

[49]   Jacoby G, Cattoir V, Hooper D, *et al. qnr* Gene nomenclature. Antimicrob Agents Chemother 2008; 52: 2297-9.

[50]   Poirel L, Liard A, Rodriguez-Martinez JM, Nordmann P. Vibrionaceae as a possible source of Qnr-like quinolone resistance determinants. J Antimicrob Chemother 2005; 56: 1118-21.

[51]   Poirel L, Rodriguez-Martinez JM, Mammeri H, Liard A, Nordmann P. Origin of plasmid-mediated quinolone resistance determinant QnrA. Antimicrob Agents Chemother 2005; 49: 3523-5.

[52]   Perichon B, Courvalin P, Galimand M. Transferable resistance to aminoglycosides by methylation of G1405 in 16S rRNA and to hydrophilic fluoroquinolones by QepA-mediated efflux in *Escherichia coli*. Antimicrob Agents Chemother 2007; 51: 2464-9.

[53]   Yamane K, Wachino J, Suzuki S, *et al.* New plasmid-mediated fluoroquinolone efflux pump, QepA, found in an *Escherichia coli* clinical isolate. Antimicrob Agents Chemother 2007; 51: 3354-60.

[54]   Poirel L, Pitout JD, Calvo L, *et al. In vivo* selection of fluoroquinolone-resistant *Escherichia coli* isolates expressing plasmid-mediated quinolone resistance and expanded-spectrum beta-lactamase. Antimicrob Agents Chemother 2006; 50: 1525-7.

[55]   Yue L, Jiang HX, Liao XP, *et al.* Prevalence of plasmid-mediated quinolone resistance *qnr* genes in poultry and swine clinical isolates of *Escherichia coli*. Vet Microbiol 2008; 132: 414-20.

[56]   Ma J, Zeng Z, Chen Z, *et al.* High prevalence of plasmid-mediated quinolone resistance determinants *qnr, aac(6')-Ib-cr*, and *qepA* among ceftiofur-resistant *Enterobacteriaceae* isolates from companion and food-producing animals. Antimicrob Agents Chemother 2009; 53: 519-24.

[57]   Piddock LJ. Clinically relevant chromosomally encoded multidrug resistance efflux pumps in bacteria. Clin Microbiol Rev 2006; 19: 382-402.

[58]   Levy SB. Active efflux mechanisms for antimicrobial resistance. Antimicrob Agents Chemother 1992; 36: 695-703.

[59]   Zhanel GG, Dueck M, Hoban DJ, *et al.* Review of macrolides and ketolides: focus on respiratory tract infections. Drugs 2001; 61: 443-98.

[60]   Paulsen IT. Multidrug efflux pumps and resistance: regulation and evolution. Curr Opin Microbiol 2003; 6: 446-51.

[61]   Folster JP, Shafer WM. Regulation of *mtrF* expression in *Neisseria gonorrhoeae* and its role in high-level antimicrobial resistance. J Bacteriol 2005; 187: 3713-20.

[62]   Nikaido H. Molecular basis of bacterial outer membrane permeability revisited. Microbiol Mol Biol Rev 2003; 67: 593-656.

[63]   Olliver A, Valle M, Chaslus-Dancla E, Cloeckaert A. Role of an *acrR* mutation in multidrug resistance of *in vitro*-selected fluoroquinolone-resistant mutants of *Salmonella* enterica serovar Typhimurium. FEMS Microbiol Lett 2004; 238: 267-72.

[64]   Tegos G, Stermitz FR, Lomovskaya O, Lewis K. Multidrug pump inhibitors uncover remarkable activity of plant antimicrobials. Antimicrob Agents Chemother 2002; 46: 3133-41.

[65]   Wolter DJ, Smith-Moland E, Goering RV, Hanson ND, Lister PD. Multidrug resistance associated with *mexXY* expression in clinical isolates of *Pseudomonas aeruginosa* from a Texas hospital. Diagn Microbiol Infect Dis 2004; 50: 43-50.

[66]   Sulavik MC, Dazer M, Miller PF. The *Salmonella* typhimurium *mar* locus: molecular and genetic analyses and assessment of its role in virulence. J Bacteriol 1997; 179: 1857-66.

[67]   Mazzariol A, Tokue Y, Kanegawa TM, Cornaglia G, Nikaido H. High-level fluoroquinolone-resistant clinical isolates of *Escherichia coli* overproduce multidrug efflux protein AcrA. Antimicrob Agents Chemother 2000; 44: 3441-3.

[68]   Mazzariol A, Zuliani J, Cornaglia G, Rossolini GM, Fontana R. AcrAB Efflux System: Expression and Contribution to Fluoroquinolone Resistance in *Klebsiella* spp. Antimicrob Agents Chemother 2002; 46: 3984-6.

[69]   Oethinger M, Podglajen I, Kern WV, Levy SB. Overexpression of the *marA* or *soxS* regulatory gene in clinical topoisomerase mutants of *Escherichia coli*. Antimicrob Agents Chemother 1998; 42: 2089-94.

[70]   Webber MA, Piddock LJ. Absence of mutations in *marRAB* or *soxRS* in *acrB*-overexpressing fluoroquinolone-resistant clinical and veterinary isolates of *Escherichia coli*. Antimicrob Agents Chemother 2001; 45: 1550-2.

[71]   Webber MA, Talukder A, Piddock LJ. Contribution of mutation at amino acid 45 of AcrR to *acrB* expression and ciprofloxacin resistance in clinical and veterinary *Escherichia coli* isolates. Antimicrob Agents Chemother 2005; 49: 4390-2.

[72]   Baucheron S, Tyler S, Boyd D, *et al.* AcrAB-TolC directs efflux-mediated multidrug resistance in *Salmonella* enterica serovar typhimurium DT104. Antimicrob Agents Chemother 2004; 48: 3729-35.

[73]   Giraud E, Cloeckaert A, Kerboeuf D, Chaslus-Dancla E. Evidence for active efflux as the primary mechanism of resistance to ciprofloxacin in *Salmonella* enterica serovar typhimurium. Antimicrob Agents Chemother 2000; 44: 1223-8.

[74]   Piddock LJ, White DG, Gensberg K, Pumbwe L, Griggs DJ. Evidence for an efflux pump mediating multiple antibiotic resistance in *Salmonella* enterica serovar Typhimurium. Antimicrob Agents Chemother 2000; 44: 3118-21.

[75]   Kong KF, Schneper L, Mathee K. Beta-lactam antibiotics: from antibiosis to resistance and bacteriology. APMIS; 118: 1-36.

[76]   Bush K, Jacoby GA, Medeiros AA. A functional classification scheme for beta-lactamases and its correlation with molecular structure. Antimicrob Agents Chemother 1995; 39: 1211-33.

[77]   Livermore DM. beta-Lactamases in laboratory and clinical resistance. Clin Microbiol Rev 1995; 8: 557-84.

[78]   Jacoby GA, Munoz-Price LS. The new beta-lactamases. N Engl J Med 2005; 352: 380-91.

[79]   Jaurin B, Normark S. Insertion of IS2 creates a novel *ampC* promoter in *Escherichia coli*. Cell 1983; 32: 809-16.

[80]   Siu LK, Lu PL, Chen JY, Lin FM, Chang SC. High-level expression of *ampC* beta-lactamase due to insertion of nucleotides between -10 and -35 promoter sequences in *Escherichia coli* clinical isolates: cases not responsive to extended-spectrum-cephalosporin treatment. Antimicrob Agents Chemother 2003; 47: 2138-44.

[81]   Caroff N, Espaze E, Berard I, Richet H, Reynaud A. Mutations in the *ampC* promoter of *Escherichia coli* isolates resistant to oxyiminocephalosporins without extended spectrum beta-lactamase production. FEMS Microbiol Lett 1999; 173: 459-65.

[82]   Kojima A, Ishii Y, Ishihara K, *et al.* Extended-spectrum-beta-lactamase-producing *Escherichia coli* strains isolated from farm animals from 1999 to 2002: report from the Japanese Veterinary Antimicrobial Resistance Monitoring Program. Antimicrob Agents Chemother 2005; 49: 3533-7.

[83]   Antunes P, Machado J, Sousa JC, Peixe L. Dissemination amongst humans and food products of animal origin of a *Salmonella* typhimurium clone expressing an integron-borne OXA-30 beta-lactamase. J Antimicrob Chemother 2004; 54: 429-34.

[84]   Boyd DA, Mulvey MR. OXA-1 is OXA-30 is OXA-1. J Antimicrob Chemother 2006; 58: 224-5.

[85]   Brinas L, Moreno MA, Teshager T, *et al.* Monitoring and characterization of extended-spectrum beta-lactamases in *Escherichia coli* strains from healthy and sick animals in Spain in 2003. Antimicrob Agents Chemother 2005; 49: 1262-4.

[86]   Antunes P, Machado J, Peixe L. Characterization of antimicrobial resistance and class 1 and 2 integrons in *Salmonella* enterica isolates from different sources in Portugal. J Antimicrob Chemother 2006; 58: 297-304.

[87]   Mulvey MR, Boyd DA, Olson AB, Doublet B, Cloeckaert A. The genetics of *Salmonella* genomic island 1. Microbes Infect 2006; 8: 1915-22.

[88]   Li XZ, Mehrotra M, Ghimire S, Adewoye L. beta-Lactam resistance and beta-lactamases in bacteria of animal origin. Vet Microbiol 2007; 121: 197-214.

[89]   Philippon A, Arlet G, Jacoby GA. Plasmid-determined AmpC-type beta-lactamases. Antimicrob Agents Chemother 2002; 46: 1-11.

[90]   Alcaine SD, Sukhnanand SS, Warnick LD, *et al.* Ceftiofur-resistant *Salmonella* strains isolated from dairy farms represent multiple widely distributed subtypes that evolved by independent horizontal gene transfer. Antimicrob Agents Chemother 2005; 49: 4061-7.

[91]   Allen KJ, Poppe C. Phenotypic and genotypic characterization of food animal isolates of *Salmonella* with reduced sensitivity to ciprofloxacin. Microb Drug Resist 2002; 8: 375-83.

[92]   Bradford PA. Extended-spectrum beta-lactamases in the 21st century: characterization, epidemiology, and detection of this important resistance threat. Clin Microbiol Rev 2001; 14: 933-51, table of contents.

[93]   Bonnet R. Growing group of extended-spectrum beta-lactamases: the CTX-M enzymes. Antimicrob Agents Chemother 2004; 48: 1-14.

[94]   Costa D, Poeta P, Brinas L, *et al.* Detection of CTX-M-1 and TEM-52 beta-lactamases in *Escherichia coli* strains from healthy pets in Portugal. J Antimicrob Chemother 2004; 54: 960-1.

[95]   Chen S, Zhao S, White DG, *et al.* Characterization of multiple-antimicrobial-resistant *salmonella* serovars isolated from retail meats. Appl Environ Microbiol 2004; 70: 1-7.

[96]   Brinas L, Moreno MA, Zarazaga M, *et al.* Detection of CMY-2, CTX-M-14, and SHV-12 beta-lactamases in *Escherichia coli* fecal-sample isolates from healthy chickens. Antimicrob Agents Chemother 2003; 47: 2056-8.

[97]   Olesen I, Hasman H, Aarestrup FM. Prevalence of beta-lactamases among ampicillin-resistant *Escherichia coli* and *Salmonella* isolated from food animals in Denmark. Microb Drug Resist 2004; 10: 334-40.

[98]   Barber M. Methicillin-resistant *staphylococci*. J Clin Pathol 1961; 14: 385-93.

[99]   Zapun A, Contreras-Martel C, Vernet T. Penicillin-binding proteins and beta-lactam resistance. FEMS Microbiol Rev 2008; 32: 361-85.

[100]   Fuda CC, Fisher JF, Mobashery S. Beta-lactam resistance in *Staphylococcus aureus*: the adaptive resistance of a plastic genome. Cell Mol Life Sci 2005; 62: 2617-33.

[101]   Alekshun MN, Levy SB. Molecular mechanisms of antibacterial multidrug resistance. Cell 2007; 128: 1037-50.

[102]   Diep BA, Gill SR, Chang RF, *et al.* Complete genome sequence of USA300, an epidemic clone of community-acquired meticillin-resistant *Staphylococcus aureus*. Lancet 2006; 367: 731-9.

[103]   Naimi TS, LeDell KH, Como-Sabetti K, *et al.* Comparison of community- and health care-associated methicillin-resistant *Staphylococcus aureus* infection. JAMA 2003; 290: 2976-84.

[104]   Fontana R, Aldegheri M, Ligozzi M, *et al.* Overproduction of a low-affinity penicillin-binding protein and high-level ampicillin resistance in *Enterococcus faecium*. Antimicrob Agents Chemother 1994; 38: 1980-3.

[105]   Klare I, Rodloff AC, Wagner J, Witte W, Hakenbeck R. Overproduction of a penicillin-binding protein is not the only mechanism of penicillin resistance in *Enterococcus faecium*. Antimicrob Agents Chemother 1992; 36: 783-7.

[106]   Laible G, Spratt BG, Hakenbeck R. Interspecies recombinational events during the evolution of altered PBP 2x genes in penicillin-resistant clinical isolates of *Streptococcus pneumoniae*. Mol Microbiol 1991; 5: 1993-2002.

[107]   Li XZ, Ma D, Livermore DM, Nikaido H. Role of efflux pump(s) in intrinsic resistance of *Pseudomonas aeruginosa*: active efflux as a contributing factor to beta-lactam resistance. Antimicrob Agents Chemother 1994; 38: 1742-52.

[108]   Nikaido H, Basina M, Nguyen V, Rosenberg EY. Multidrug efflux pump AcrAB of *Salmonella* typhimurium excretes only those beta-lactam antibiotics containing lipophilic side chains. J Bacteriol 1998; 180: 4686-92.

[109]   Kim BJ, Lee KH, Yun YJ, *et al.* Simultaneous identification of rifampin-resistant *Mycobacterium tuberculosis* and nontuberculous *mycobacteria* by polymerase chain reaction-single strand conformation polymorphism and sequence analysis of the RNA polymerase gene (*rpoB*). J Microbiol Methods 2004; 58: 111-8.

[110]   Baucheron S, Imberechts H, Chaslus-Dancla E, Cloeckaert A. The AcrB multidrug transporter plays a major role in high-level fluoroquinolone resistance in *Salmonella* enterica serovar typhimurium phage type DT204. Microb Drug Resist 2002; 8: 281-9.

[111]   Olliver A, Valle M, Chaslus-Dancla E, Cloeckaert A. Overexpression of the multidrug efflux operon *acrEF* by insertional activation with IS1 or IS10 elements in *Salmonella* enterica serovar typhimurium DT204 acrB mutants selected with fluoroquinolones. Antimicrob Agents Chemother 2005; 49: 289-301.

[112]   Chopra I, Roberts M. Tetracycline antibiotics: mode of action, applications, molecular biology, and epidemiology of bacterial resistance. Microbiol Mol Biol Rev 2001; 65: 232-60 ; second page, table of contents.

[113]   Yamaguchi A, Adachi K, Akasaka T, Ono N, Sawai T. Metal-tetracycline/H+ antiporter of *Escherichia coli* encoded by a transposon Tn10. Histidine 257 plays an essential role in H+ translocation. J Biol Chem 1991; 266: 6045-51.

[114]   Jones CS, Osborne DJ, Stanley J. Enterobacterial tetracycline resistance in relation to plasmid incompatibility. Mol Cell Probes 1992; 6: 313-7.

[115]   Mendez B, Tachibana C, Levy SB. Heterogeneity of tetracycline resistance determinants. Plasmid 1980; 3: 99-108.

[116]   Mazel D. Integrons: agents of bacterial evolution. Nat Rev Microbiol 2006; 4: 608-20.

[117]   Roberts MC. Tetracycline resistance in *Peptostreptococcus* species. Antimicrob Agents Chemother 1991; 35: 1682-4.

[118]   Connell SR, Tracz DM, Nierhaus KH, Taylor DE. Ribosomal protection proteins and their mechanism of tetracycline resistance. Antimicrob Agents Chemother 2003; 47: 3675-81.

[119]   Sanchez-Pescador R, Brown JT, Roberts M, Urdea MS. Homology of the TetM with translational elongation factors: implications for potential modes of *tetM*-conferred tetracycline resistance. Nucleic Acids Res 1988; 16: 1218.

[120]   Taylor DE, Chau A. Tetracycline resistance mediated by ribosomal protection. Antimicrob Agents Chemother 1996; 40: 1-5.

[121] Dantley KA, Dannelly HK, Burdett V. Binding interaction between Tet(M) and the ribosome: requirements for binding. J Bacteriol 1998; 180: 4089-92.

[122] Sloan J, McMurry LM, Lyras D, Levy SB, Rood JI. The *Clostridium perfringens* Tet P determinant comprises two overlapping genes: *tetA(P),* which mediates active tetracycline efflux, and *tetB(P),* which is related to the ribosomal protection family of tetracycline-resistance determinants. Mol Microbiol 1994; 11: 403-15.

[123] Speer BS, Bedzyk L, Salyers AA. Evidence that a novel tetracycline resistance gene found on two *Bacteroides* transposons encodes an NADP-requiring oxidoreductase. J Bacteriol 1991; 173: 176-83.

[124] Moore IF, Hughes DW, Wright GD. Tigecycline is modified by the flavin-dependent monooxygenase TetX. Biochemistry 2005; 44: 11829-35.

[125] Kordick DL, Papich MG, Breitschwerdt EB. Efficacy of enrofloxacin or doxycycline for treatment of *Bartonella henselae* or *Bartonella clarridgeiae* infection in cats. Antimicrob Agents Chemother 1997; 41: 2448-55.

[126] Stokstad EL, Jukes TH, *et al.* The multiple nature of the animal protein factor. J Biol Chem 1949; 180: 647-54.

[127] Chopra I, Hawkey PM, Hinton M. Tetracyclines, molecular and clinical aspects. J Antimicrob Chemother 1992; 29: 245-77.

[128] Levy SB, FitzGerald GB, Macone AB. Changes in intestinal flora of farm personnel after introduction of a tetracycline-supplemented feed on a farm. N Engl J Med 1976; 295: 583-8.

[129] Linton AH. Antibiotic-resistant bacteria in animal husbandry. Br Med Bull 1984; 40: 91-5.

[130] Manie T, Khan S, Brozel VS, Veith WJ, Gouws PA. Antimicrobial resistance of bacteria isolated from slaughtered and retail chickens in South Africa. Lett Appl Microbiol 1998; 26: 253-8.

[131] Witte W. Medical consequences of antibiotic use in agriculture. Science 1998; 279: 996-7.

[132] Kobland JD, Gale GO, Gustafson RH, Simkins KL. Comparison of therapeutic versus subtherapeutic levels of chlortetracycline in the diet for selection of resistant *salmonella* in experimentally challenged chickens. Poult Sci 1987; 66: 1129-37.

[133] Fey PD, Safranek TJ, Rupp ME, *et al.* Ceftriaxone-resistant *salmonella* infection acquired by a child from cattle. N Engl J Med 2000; 342: 1242-9.

[134] Molbak K, Baggesen DL, Aarestrup FM, *et al.* An outbreak of multidrug-resistant, quinolone-resistant *Salmonella* enterica serotype typhimurium DT104. N Engl J Med 1999; 341: 1420-5.

[135] Skold O. Resistance to trimethoprim and sulfonamides. Vet Res 2001; 32: 261-73.

[136] Huovinen P. Resistance to trimethoprim-sulfamethoxazole. Clin Infect Dis 2001; 32: 1608-14.

[137] Dale GE, Broger C, D'Arcy A, *et al.* A single amino acid substitution in *Staphylococcus aureus* dihydrofolate reductase determines trimethoprim resistance. J Mol Biol 1997; 266: 23-30.

[138] Pikis A, Donkersloot JA, Rodriguez WJ, Keith JM. A conservative amino acid mutation in the chromosome-encoded dihydrofolate reductase confers trimethoprim resistance in *Streptococcus pneumoniae*. J Infect Dis 1998; 178: 700-6.

[139] de Groot R, Sluijter M, de Bruyn A, *et al.* Genetic characterization of trimethoprim resistance in *Haemophilus influenzae*. Antimicrob Agents Chemother 1996; 40: 2131-6.

[140] Radstrom P, Swedberg G. RSF1010 and a conjugative plasmid contain sulII, one of two known genes for plasmid-borne sulfonamide resistance dihydropteroate synthase. Antimicrob Agents Chemother 1988; 32: 1684-92.

[141] Schwarz S, Chaslus-Dancla E. Use of antimicrobials in veterinary medicine and mechanisms of resistance. Vet Res 2001; 32: 201-25.

[142] Jansson C, Skold O. Appearance of a new trimethoprim resistance gene, *dhfrIX*, in *Escherichia coli* from swine. Antimicrob Agents Chemother 1991; 35: 1891-9.

[143] Jansson C, Franklin A, Skold O. Spread of a newly found trimethoprim resistance gene, *dhfrIX*, among porcine isolates and human pathogens. Antimicrob Agents Chemother 1992; 36: 2704-8.

[144] Shaheen BW, Boothe DM, Oyarzabal OA. The role of Class 1 and 2 integrons in mediating antimicrobial resistance among canine and feline clinical *Escherichia coli* isolates from the US. Vet Microbiol 2010; (in Press).

[145] Davies J, Wright GD. Bacterial resistance to aminoglycoside antibiotics. Trends Microbiol 1997; 5: 234-40.

[146] Mingeot-Leclercq MP, Glupczynski Y, Tulkens PM. Aminoglycosides: activity and resistance. Antimicrob Agents Chemother 1999; 43: 727-37.

[147] Shaw KJ, Rather PN, Hare RS, Miller GH. Molecular genetics of aminoglycoside resistance genes and familial relationships of the aminoglycoside-modifying enzymes. Microbiol Rev 1993; 57: 138-63.

[148] Wright GD. Aminoglycoside-modifying enzymes. Curr Opin Microbiol 1999; 2: 499-503.

[149] Chaslus-Dancla E, Martel JL, Carlier C, Lafont JP, Courvalin P. Emergence of aminoglycoside 3-N-acetyltransferase IV in *Escherichia coli* and *Salmonella* typhimurium isolated from animals in France. Antimicrob Agents Chemother 1986; 29: 239-43.

[150] Chaslus-Dancla E, Pohl P, Meurisse M, Marin M, Lafont JP. High genetic homology between plasmids of human and animal origins conferring resistance to the aminoglycosides gentamicin and apramycin. Antimicrob Agents Chemother 1991; 35: 590-3.

[151] Lange CC, Werckenthin C, Schwarz S. Molecular analysis of the plasmid-borne *aacA/aphD* resistance gene region of coagulase-negative *staphylococci* from chickens. J Antimicrob Chemother 2003; 51: 1397-401.

[152] Lyon BR, Skurray R. Antimicrobial resistance of *Staphylococcus aureus*: genetic basis. Microbiol Rev 1987; 51: 88-134.

[153] Rouch DA, Byrne ME, Kong YC, Skurray RA. The *aacA-aphD* gentamicin and kanamycin resistance determinant of Tn4001 from *Staphylococcus aureus*: expression and nucleotide sequence analysis. J Gen Microbiol 1987; 133: 3039-52.

[154] Recchia GD, Hall RM. Gene cassettes: a new class of mobile element. Microbiology 1995; 141 (Pt 12): 3015-27.

[155] Poirel L, Lambert T, Turkoglu S, *et al.* Characterization of Class 1 integrons from *Pseudomonas aeruginosa* that contain the *bla(VIM-2)* carbapenem-hydrolyzing beta-lactamase gene and of two novel aminoglycoside resistance gene cassettes. Antimicrob Agents Chemother 2001; 45: 546-52.

[156] Rosenberg EY, Ma D, Nikaido H. *AcrD* of *Escherichia coli* is an aminoglycoside efflux pump. J Bacteriol 2000; 182: 1754-6.

[157] Aires JR, Kohler T, Nikaido H, Plesiat P. Involvement of an active efflux system in the natural resistance of *Pseudomonas aeruginosa* to aminoglycosides. Antimicrob Agents Chemother 1999; 43: 2624-8.

[158] Schwarz S, Kehrenberg C, Doublet B, Cloeckaert A. Molecular basis of bacterial resistance to chloramphenicol and florfenicol. FEMS Microbiol Rev 2004; 28: 519-42.

[159] Murray IA, Shaw WV. O-Acetyltransferases for chloramphenicol and other natural products. Antimicrob Agents Chemother 1997; 41: 1-6.

[160] Shaw WV. Chloramphenicol acetyltransferase: enzymology and molecular biology. CRC Crit Rev Biochem 1983; 14: 1-46.

[161] Murray IA, Hawkins AR, Keyte JW, Shaw WV. Nucleotide sequence analysis and overexpression of the gene encoding a type III chloramphenicol acetyltransferase. Biochem J 1988; 252: 173-9.

[162] Murray IA, Martinez-Suarez JV, Close TJ, Shaw WV. Nucleotide sequences of genes encoding the type II chloramphenicol acetyltransferases of *Escherichia coli* and *Haemophilus influenzae*, which are sensitive to inhibition by thiol-reactive reagents. Biochem J 1990; 272: 505-10.

[163] Alton NK, Vapnek D. Nucleotide sequence analysis of the chloramphenicol resistance transposon Tn9. Nature 1979; 282: 864-9.

[164] Elisha BG, Steyn LM. Identification of an *Acinetobacter baumannii* gene region with sequence and organizational similarity to Tn2670. Plasmid 1991; 25: 96-104.

[165] Kim E, Aoki T. The structure of the chloramphenicol resistance gene on a transferable R plasmid from the fish pathogen, *Pasteurella piscicida*. Microbiol Immunol 1993; 37: 705-12.

[166] Roberts M, Corney A, Shaw WV. Molecular characterization of three chloramphenicol acetyltransferases isolated from *Haemophilus influenzae*. J Bacteriol 1982; 151: 737-41.

[167] Paulsen IT, Brown MH, Skurray RA. Proton-dependent multidrug efflux systems. Microbiol Rev 1996; 60: 575-608.

[168] McMurry LM, George AM, Levy SB. Active efflux of chloramphenicol in susceptible *Escherichia coli* strains and in multiple-antibiotic-resistant (Mar) mutants. Antimicrob Agents Chemother 1994; 38: 542-6.

[169] Poole K, Srikumar R. Multidrug efflux in *Pseudomonas aeruginosa*: components, mechanisms and clinical significance. Curr Top Med Chem 2001; 1: 59-71.

[170] Hancock RE, Brinkman FS. Function of pseudomonas porins in uptake and efflux. Annu Rev Microbiol 2002; 56: 17-38.

[171] Simonet V, Mallea M, Pages JM. Substitutions in the eyelet region disrupt cefepime diffusion through the *Escherichia coli* OmpF channel. Antimicrob Agents Chemother 2000; 44: 311-5.

[172] Martinez-Martinez L, Hernandez-Alles S, Alberti S, *et al. In vivo* selection of porin-deficient mutants of *Klebsiella pneumoniae* with increased resistance to cefoxitin and expanded-spectrum-cephalosporins. Antimicrob Agents Chemother 1996; 40: 342-8.

[173] Cohen SP, McMurry LM, Hooper DC, Wolfson JS, Levy SB. Cross-resistance to fluoroquinolones in multiple-antibiotic-resistant (Mar) *Escherichia coli* selected by tetracycline or chloramphenicol: decreased drug accumulation associated with membrane changes in addition to OmpF reduction. Antimicrob Agents Chemother 1989; 33: 1318-25.

[174] Institute CaLS. Performance Standards for Antimicrobial Disk and Dilution Susceptibility Testing for Bacteria Isolated From Animals; Approved Standard-Third Edition . CLSI Document M31-A3 2008; 28.

[175] Kaltenboeck B, Wang C. Advances in real-time PCR: application to clinical laboratory diagnostics. Adv Clin Chem 2005; 40: 219-59.

[176] Saiki RK, Gelfand DH, Stoffel S, *et al.* Primer-directed enzymatic amplification of DNA with a thermostable DNA polymerase. Science 1988; 239: 487-91.

[177] Mullis KB. The unusual origin of the polymerase chain reaction. Sci Am 1990; 262: 56-61, 4-5.

[178] Cockerill FR, 3rd. Genetic methods for assessing antimicrobial resistance. Antimicrob Agents Chemother 1999; 43: 199-212.

[179] Zirnstein G, Li Y, Swaminathan B, Angulo F. Ciprofloxacin resistance in *Campylobacter jejuni* isolates: detection of *gyrA* resistance mutations by mismatch amplification mutation assay PCR and DNA sequence analysis. J Clin Microbiol 1999; 37: 3276-80.

[180] Arens M. Methods for subtyping and molecular comparison of human viral genomes. Clin Microbiol Rev 1999; 12: 612-26.

[181] Fluit AC, Visser MR, Schmitz FJ. Molecular detection of antimicrobial resistance. Clin Microbiol Rev 2001; 14: 836-71, table of contents.

[182] Geha DJ, Uhl JR, Gustaferro CA, Persing DH. Multiplex PCR for identification of methicillin-resistant *staphylococci* in the clinical laboratory. J Clin Microbiol 1994; 32: 1768-72.

[183] Wallet F, Roussel-Delvallez M, Courcol RJ. Choice of a routine method for detecting methicillin-resistance in *staphylococci*. J Antimicrob Chemother 1996; 37: 901-9.

[184] Kitagawa Y, Ueda M, Ando N, *et al.* Rapid diagnosis of methicillin-resistant *Staphylococcus aureus* bacteremia by nested polymerase chain reaction. Ann Surg 1996; 224: 665-71.

[185] Englen MD, Fedorka-Cray PJ. Evaluation of a commercial diagnostic PCR for the identification of *Campylobacter jejuni* and *Campylobacter coli*. Lett Appl Microbiol 2002; 35: 353-6.

[186] Liu C, Song Y, McTeague M, *et al.* Rapid identification of the species of the *Bacteroides fragilis* group by multiplex PCR assays using group- and species-specific primers. FEMS Microbiol Lett 2003; 222: 9-16.

[187] McClure JA, Conly JM, Lau V, *et al.* Novel multiplex PCR assay for detection of the staphylococcal virulence marker Panton-Valentine leukocidin genes and simultaneous discrimination of methicillin-susceptible from -resistant *staphylococci*. J Clin Microbiol 2006; 44: 1141-4.

[188] Zhang K, Sparling J, Chow BL, *et al.* New quadriplex PCR assay for detection of methicillin and mupirocin resistance and simultaneous discrimination of *Staphylococcus aureus* from coagulase-negative *staphylococci*. J Clin Microbiol 2004; 42: 4947-55.

[189] Kolbert CP, Arruda J, Varga-Delmore P, *et al.* Branched-DNA assay for detection of the *mecA* gene in oxacillin-resistant and oxacillin-sensitive *staphylococci*. J Clin Microbiol 1998; 36: 2640-4.

[190] Mulder JG. Comparison of disk diffusion, the E test, and detection of *mecA* for determination of methicillin resistance in coagulase-negative *staphylococci*. Eur J Clin Microbiol Infect Dis 1996; 15: 567-73.

[191] Granger K, Rundell MS, Pingle MR, *et al.* Multiplex PCR-Ligation Detection Reaction assay for the simultaneous detection of drug resistance and toxin genes from *Staphylococcus aureus*, *Enterococcus faecalis* and *Enterococcus faecium*. J Clin Microbiol 2009.

[192] Qiang YZ, Qin T, Fu W, *et al.* Use of a rapid mismatch PCR method to detect *gyrA* and *parC* mutations in ciprofloxacin-resistant clinical isolates of *Escherichia coli*. J Antimicrob Chemother 2002; 49: 549-52.

[193] Zuna J, Muzikova K, Madzo J, Krejci O, Trka J. Temperature non-homogeneity in rapid airflow-based cycler significantly affects real-time PCR. Biotechniques 2002; 33: 508, 10, 12.

[194] Brow MA, Oldenburg MC, Lyamichev V, *et al.* Differentiation of bacterial 16S rRNA genes and intergenic regions and *Mycobacterium tuberculosis* katG genes by structure-specific endonuclease cleavage. J Clin Microbiol 1996; 34: 3129-37.

[195] Cockerill FR, 3rd, Uhl JR, Temesgen Z, *et al.* Rapid identification of a point mutation of the *Mycobacterium tuberculosis* catalase-peroxidase (katG) gene associated with isoniazid resistance. J Infect Dis 1995; 171: 240-5.

[196] Heym B, Alzari PM, Honore N, Cole ST. Missense mutations in the catalase-peroxidase gene, *katG*, are associated with isoniazid resistance in *Mycobacterium tuberculosis*. Mol Microbiol 1995; 15: 235-45.

[197] Marttila HJ, Soini H, Huovinen P, Viljanen MK. *katG* mutations in isoniazid-resistant *Mycobacterium tuberculosis* isolates recovered from Finnish patients. Antimicrob Agents Chemother 1996; 40: 2187-9.

[198] Nachamkin I, Kang C, Weinstein MP. Detection of resistance to isoniazid, rifampin, and streptomycin in clinical isolates of *Mycobacterium tuberculosis* by molecular methods. Clin Infect Dis 1997; 24: 894-900.

[199] Rouse DA, Li Z, Bai GH, Morris SL. Characterization of the *katG* and *inhA* genes of isoniazid-resistant clinical isolates of *Mycobacterium tuberculosis.* Antimicrob Agents Chemother 1995; 39: 2472-7.

[200] Varela G, Gonzalez S, Gadea P, *et al.* Prevalence and dissemination of the Ser315Thr substitution within the KatG enzyme in isoniazid-resistant strains of *Mycobacterium tuberculosis* isolated in Uruguay. J Med Microbiol 2008; 57: 1518-22.

[201] Caws M, Tho DQ, Duy PM, *et al.* PCR-restriction fragment length polymorphism for rapid, low-cost identification of isoniazid-resistant *Mycobacterium tuberculosis.* J Clin Microbiol 2007; 45: 1789-93.

[202] Ip M, Chau SS, Chi F, Qi A, Lai RW. Rapid screening of fluoroquinolone resistance determinants in *Streptococcus pneumoniae* by PCR-restriction fragment length polymorphism and single-strand conformational polymorphism. J Clin Microbiol 2006; 44: 970-5.

[203] San Martin B, Lapierre L, Toro C, *et al.* Isolation and molecular characterization of quinolone resistant *Salmonella* spp. from poultry farms. Vet Microbiol 2005; 110: 239-44.

[204] Alonso R, Galimand M, Courvalin P. An extended PCR-RFLP assay for detection of *parC, parE* and *gyrA* mutations in fluoroquinolone-resistant *Streptococcus pneumoniae.* J Antimicrob Chemother 2004; 53: 682-3.

[205] Huang TM, Chang YF, Chang CF. Detection of mutations in the *gyrA* gene and class I integron from quinolone-resistant *Salmonella* enterica serovar Choleraesuis isolates in Taiwan. Vet Microbiol 2004; 100: 247-54.

[206] Alonso R, Mateo E, Girbau C, *et al.* PCR-restriction fragment length polymorphism assay for detection of *gyrA* mutations associated with fluoroquinolone resistance in *Campylobacter coli.* Antimicrob Agents Chemother 2004; 48: 4886-8.

[207] Takahashi H, Kikuchi T, Shoji S, *et al.* Characterization of *gyrA, gyrB, grlA* and *grlB* mutations in fluoroquinolone-resistant clinical isolates of *Staphylococcus aureus.* J Antimicrob Chemother 1998; 41: 49-57.

[208] Ouabdesselam S, Hooper DC, Tankovic J, Soussy CJ. Detection of *gyrA* and *gyrB* mutations in quinolone-resistant clinical isolates of *Escherichia coli* by single-strand conformational polymorphism analysis and determination of levels of resistance conferred by two different single *gyrA* mutations. Antimicrob Agents Chemother 1995; 39: 1667-70.

[209] Santos OC, Barros EM, Brito MA, *et al.* Identification of coagulase-negative *staphylococci* from bovine mastitis using RFLP-PCR of the *groEL* gene. Vet Microbiol 2008; 130: 134-40.

[210] Wittwer M, Keller J, Wassenaar TM, *et al.* Genetic diversity and antibiotic resistance patterns in a *campylobacter* population isolated from poultry farms in Switzerland. Appl Environ Microbiol 2005; 71: 2840-7.

[211] Wang GQ, Wu CM, Du XD, *et al.* Characterization of integrons-mediated antimicrobial resistance among *Escherichia coli* strains isolated from bovine mastitis. Vet Microbiol 2008; 127: 73-8.

[212] Alonso R, Fernandez-Aranguiz A, Colom K, Cisterna R. Non-radioactive PCR-SSCP with a single PCR step for detection of inhibitor resistant beta-lactamases in *Escherichia coli.* J Microbiol Methods 2002; 50: 85-90.

[213] Temesgen Z, Satoh K, Uhl JR, Kline BC, Cockerill FR, 3rd. Use of polymerase chain reaction single-strand conformation polymorphism (PCR-SSCP) analysis to detect a point mutation in the catalase-peroxidase gene (*katG*) of *Mycobacterium tuberculosis.* Mol Cell Probes 1997; 11: 59-63.

[214] Cardoso RF, Cooksey RC, Morlock GP, *et al.* Screening and characterization of mutations in isoniazid-resistant *Mycobacterium tuberculosis* isolates obtained in Brazil. Antimicrob Agents Chemother 2004; 48: 3373-81.

[215] Whelen AC, Felmlee TA, Hunt JM, *et al.* Direct genotypic detection of *Mycobacterium tuberculosis* rifampin resistance in clinical specimens by using single-tube heminested PCR. J Clin Microbiol 1995; 33: 556-61.

[216] Sheng J, Li J, Sheng G, *et al.* Characterization of *rpoB* mutations associated with rifampin resistance in *Mycobacterium tuberculosis* from eastern China. J Appl Microbiol 2008; 105: 904-11.

[217] Sreevatsan S, Stockbauer KE, Pan X, *et al.* Ethambutol resistance in *Mycobacterium tuberculosis*: critical role of *embB* mutations. Antimicrob Agents Chemother 1997; 41: 1677-81.

[218] Takiff HE, Salazar L, Guerrero C, *et al.* Cloning and nucleotide sequence of *Mycobacterium tuberculosis gyrA* and *gyrB* genes and detection of quinolone resistance mutations. Antimicrob Agents Chemother 1994; 38: 773-80.

[219] Hayashi K. PCR-SSCP: a simple and sensitive method for detection of mutations in the genomic DNA. PCR Methods Appl 1991; 1: 34-8.

[220] Orita M, Iwahana H, Kanazawa H, Hayashi K, Sekiya T. Detection of polymorphisms of human DNA by gel electrophoresis as single-strand conformation polymorphisms. Proc Natl Acad Sci USA 1989; 86: 2766-70.

[221] Sarkar G, Yoon HS, Sommer SS. Dideoxy fingerprinting (ddE): a rapid and efficient screen for the presence of mutations. Genomics 1992; 13: 441-3.

[222] Telenti A, Imboden P, Marchesi F, Schmidheini T, Bodmer T. Direct, automated detection of rifampin-resistant *Mycobacterium tuberculosis* by polymerase chain reaction and single-strand conformation polymorphism analysis. Antimicrob Agents Chemother 1993; 37: 2054-8.

[223] Kim BJ, Lee KH, Park BN, *et al.* Detection of rifampin-resistant *Mycobacterium tuberculosis* in sputa by nested PCR-linked single-strand conformation polymorphism and DNA sequencing. J Clin Microbiol 2001; 39: 2610-7.

[224] Cheng X, Zhang J, Yang L, *et al.* A new Multi-PCR-SSCP assay for simultaneous detection of isoniazid and rifampin resistance in *Mycobacterium tuberculosis.* J Microbiol Methods 2007; 70: 301-5.

[225] Sheen P, Mendez M, Gilman RH, *et al.* Sputum PCR-single-strand conformational polymorphism test for same-day detection of pyrazinamide resistance in tuberculosis patients. J Clin Microbiol 2009; 47: 2937-43.

[226] Olive DM, Bean P. Principles and applications of methods for DNA-based typing of microbial organisms. J Clin Microbiol 1999; 37: 1661-9.

[227] Maddox LO, Li P, Bennett A, Descartes M, Thompson JN. Comparison of SSCP analysis and CFLP analysis for mutation detection in the human iduronate 2-sulfatase gene. Biochem Mol Biol Int 1997; 43: 1163-71.

[228] Sreevatsan S, Bookout JB, Ringpis FM, *et al.* Comparative evaluation of cleavase fragment length polymorphism with PCR-SSCP and PCR-RFLP to detect antimicrobial agent resistance in *Mycobacterium tuberculosis.* Mol Diagn 1998; 3: 81-91.

[229] Marshall DJ, Heisler LM, Lyamichev V, *et al.* Determination of hepatitis C virus genotypes in the United States by cleavase fragment length polymorphism analysis. J Clin Microbiol 1997; 35: 3156-62.

[230] Sultan Z, Nahar S, Wretlind B, Lindback E, Rahman M. Comparison of mismatch amplification mutation assay with DNA sequencing for characterization of fluoroquinolone resistance in *Neisseria gonorrhoeae.* J Clin Microbiol 2004; 42: 591-4.

[231] Lin CC, Chen TH, Wang YC, *et al.* Analysis of ciprofloxacin-resistant *Salmonella* strains from swine, chicken, and their carcasses in Taiwan and detection of *parC* resistance mutations by a mismatch amplification mutation assay PCR. J Food Prot 2009; 72: 14-20.

[232] Guevremont E, Nadeau E, Sirois M, Quessy S. Antimicrobial susceptibilities of thermophilic *Campylobacter* from humans, swine, and chicken broilers. Can J Vet Res 2006; 70: 81-6.

[233] Holland PM, Abramson RD, Watson R, Gelfand DH. Detection of specific polymerase chain reaction product by utilizing the 5'----3' exonuclease activity of Thermus aquaticus DNA polymerase. Proc Natl Acad Sci USA 1991; 88: 7276-80.

[234] Heid CA, Stevens J, Livak KJ, Williams PM. Real time quantitative PCR. Genome Res 1996; 6: 986-94.

[235] Simon A, Labalette P, Ordinaire I, *et al.* Use of fluorescence resonance energy transfer hybridization probes to evaluate quantitative real-time PCR for diagnosis of ocular toxoplasmosis. J Clin Microbiol 2004; 42: 3681-5.

[236] Huang J, DeGraves FJ, Gao D, *et al.* Quantitative detection of *Chlamydia* spp. by fluorescent PCR in the LightCycler. Biotechniques 2001; 30: 150-7.

[237] Fang H, Hedin G. Rapid screening and identification of methicillin-resistant *Staphylococcus aureus* from clinical samples by selective-broth and real-time PCR assay. J Clin Microbiol 2003; 41: 2894-9.

[238] Morsczeck C, Langendorfer D, Schierholz JM. A quantitative real-time PCR assay for the detection of tetR of Tn10 in *Escherichia coli* using SYBR Green and the Opticon. J Biochem Biophys Methods 2004; 59: 217-27.

[239] Reinert RR, Franken C, van der Linden M, *et al.* Molecular characterisation of macrolide resistance mechanisms of *Streptococcus pneumoniae* and *Streptococcus pyogenes* isolated in Germany, 2002-2003. Int J Antimicrob Agents 2004; 24: 43-7.

[240] Lindler LE, Fan W, Jahan N. Detection of ciprofloxacin-resistant *Yersinia pestis* by fluorogenic PCR using the LightCycler. J Clin Microbiol 2001; 39: 3649-55.

[241] Siedner MJ, Pandori M, Castro L, *et al.* Real-time PCR assay for detection of quinolone-resistant *Neisseria gonorrhoeae* in urine samples. J Clin Microbiol 2007; 45: 1250-4.

[242] Carattoli A, Dionisi A, Luzzi I. Use of a LightCycler gyrA mutation assay for identification of ciprofloxacin-resistant *Campylobacter coli.* FEMS Microbiol Lett 2002; 214: 87-93.

[243] Grisold AJ, Leitner E, Muhlbauer G, Marth E, Kessler HH. Detection of methicillin-resistant *Staphylococcus aureus* and simultaneous confirmation by automated nucleic acid extraction and real-time PCR. J Clin Microbiol 2002; 40: 2392-7.

[244] Saribas Z, Yurdakul P, Alp A, Gunalp A. Use of fluorescence resonance energy transfer for rapid detection of isoniazid resistance in *Mycobacterium tuberculosis* clinical isolates. Int J Tuberc Lung Dis 2005; 9: 181-7.

[245] Palladino S, Kay ID, Costa AM, Lambert EJ, Flexman JP. Real-time PCR for the rapid detection of *vanA* and *vanB* genes. Diagn Microbiol Infect Dis 2003; 45: 81-4.

[246] Lapierre P, Huletsky A, Fortin V, *et al*. Real-time PCR assay for detection of fluoroquinolone resistance associated with *grlA* mutations in *Staphylococcus aureus*. J Clin Microbiol 2003; 41: 3246-51.

[247] Huletsky A, Giroux R, Rossbach V, *et al*. New real-time PCR assay for rapid detection of methicillin-resistant *Staphylococcus aureus* directly from specimens containing a mixture of *staphylococci*. J Clin Microbiol 2004; 42: 1875-84.

[248] Walker RA, Saunders N, Lawson AJ, *et al*. Use of a LightCycler *gyrA* mutation assay for rapid identification of mutations conferring decreased susceptibility to ciprofloxacin in multiresistant *Salmonella* enterica serotype Typhimurium DT104 isolates. J Clin Microbiol 2001; 39: 1443-8.

[249] Gibson UE, Heid CA, Williams PM. A novel method for real time quantitative RT-PCR. Genome Res 1996; 6: 995-1001.

[250] Tan TY, Corden S, Barnes R, Cookson B. Rapid identification of methicillin-resistant *Staphylococcus aureus* from positive blood cultures by real-time fluorescence PCR. J Clin Microbiol 2001; 39: 4529-31.

[251] Wada T, Maeda S, Tamaru A, *et al*. Dual-probe assay for rapid detection of drug-resistant *Mycobacterium tuberculosis* by real-time PCR. J Clin Microbiol 2004; 42: 5277-85.

[252] Mohn SC, Ulvik A, Jureen R, *et al*. Duplex real-time PCR assay for rapid detection of ampicillin-resistant *Enterococcus faecium*. Antimicrob Agents Chemother 2004; 48: 556-60.

[253] Wilson DL, Abner SR, Newman TC, Mansfield LS, Linz JE. Identification of ciprofloxacin-resistant *Campylobacter jejuni* by use of a fluorogenic PCR assay. J Clin Microbiol 2000; 38: 3971-8.

[254] Miller MB, Tang YW. Basic concepts of microarrays and potential applications in clinical microbiology. Clin Microbiol Rev 2009; 22: 611-33.

[255] Monecke S, Berger-Bachi B, Coombs G, *et al*. Comparative genomics and DNA array-based genotyping of pandemic *Staphylococcus aureus* strains encoding Panton-Valentine leukocidin. Clin Microbiol Infect 2007; 13: 236-49.

[256] Perreten V, Vorlet-Fawer L, Slickers P, *et al*. Microarray-based detection of 90 antibiotic resistance genes of gram-positive bacteria. J Clin Microbiol 2005; 43: 2291-302.

[257] Zhu LX, Zhang ZW, Liang D, *et al*. Multiplex asymmetric PCR-based oligonucleotide microarray for detection of drug resistance genes containing single mutations in *Enterobacteriaceae*. Antimicrob Agents Chemother 2007; 51: 3707-13.

[258] Zhu LX, Zhang ZW, Wang C, *et al*. Use of a DNA microarray for simultaneous detection of antibiotic resistance genes among staphylococcal clinical isolates. J Clin Microbiol 2007; 45: 3514-21.

[259] Vora GJ, Meador CE, Bird MM, *et al*. Microarray-based detection of genetic heterogeneity, antimicrobial resistance, and the viable but nonculturable state in human pathogenic *Vibrio* spp. Proc Natl Acad Sci USA 2005; 102: 19109-14.

[260] Sougakoff W, Rodrigue M, Truffot-Pernot C, *et al*. Use of a high-density DNA probe array for detecting mutations involved in rifampicin resistance in *Mycobacterium tuberculosis*. Clin Microbiol Infect 2004; 10: 289-94.

[261] Kadlec K, Ehricht R, Monecke S, *et al*. Diversity of antimicrobial resistance pheno- and genotypes of methicillin-resistant *Staphylococcus aureus* ST398 from diseased swine. J Antimicrob Chemother 2009; 64: 1156-64.

[262] Zhang AY, Wang HN, Tian GB, *et al*. Phenotypic and genotypic characterisation of antimicrobial resistance in faecal bacteria from 30 Giant pandas. Int J Antimicrob Agents 2009; 33: 456-60.

[263] Diarra MS, Silversides FG, Diarrassouba F, *et al*. Impact of feed supplementation with antimicrobial agents on growth performance of broiler chickens, *Clostridium perfringens* and *enterococcus* counts, and antibiotic resistance phenotypes and distribution of antimicrobial resistance determinants in *Escherichia coli* isolates. Appl Environ Microbiol 2007; 73: 6566-76.

[264] Aragon LM, Navarro F, Heiser V, *et al*. Rapid detection of specific gene mutations associated with isoniazid or rifampicin resistance in *Mycobacterium tuberculosis* clinical isolates using non-fluorescent low-density DNA microarrays. J Antimicrob Chemother 2006; 57: 825-31.

[265] Gibson JS, Cobbold RN, Heisig P, *et al*. Identification of Qnr and AAC(6')-1b-cr plasmid-mediated fluoroquinolone resistance determinants in multidrug-resistant *Enterobacter* spp. isolated from extraintestinal infections in companion animals. Vet Microbiol 2009.

[266] Strommenger B, Kehrenberg C, Kettlitz C, *et al*. Molecular characterization of methicillin-resistant *Staphylococcus aureus* strains from pet animals and their relationship to human isolates. J Antimicrob Chemother 2006; 57: 461-5.

[267] Lee JH. Methicillin (Oxacillin)-resistant *Staphylococcus aureus* strains isolated from major food animals and their potential transmission to humans. Appl Environ Microbiol 2003; 69: 6489-94.

[268] Diarrassouba F, Diarra MS, Bach S, *et al.* Antibiotic resistance and virulence genes in commensal *Escherichia coli* and *Salmonella* isolates from commercial broiler chicken farms. J Food Prot 2007; 70: 1316-27.

[269] Ahmed AM, Younis EE, Osman SA, *et al.* Genetic analysis of antimicrobial resistance in *Escherichia coli* isolated from diarrheic neonatal calves. Vet Microbiol 2009; 136: 397-402.

[270] Gow SP, Waldner CL, Harel J, Boerlin P. Associations between antimicrobial resistance genes in fecal generic *Escherichia coli* isolates from cow-calf herds in western Canada. Appl Environ Microbiol 2008; 74: 3658-66.

[271] Meemken D, Blaha T, Tegeler R, *et al.* Livestock Associated Methicillin-Resistant *Staphylococcus aureus* (LaMRSA) Isolated from Lesions of Pigs at Necropsy in Northwest Germany Between 2004 and 2007. Zoonoses Public Health 2009.

[272] Mirzaagha P, Louie M, Read RR, *et al.* Characterization of tetracycline- and ampicillin-resistant *Escherichia coli* isolated from the feces of feedlot cattle over the feeding period. Can J Microbiol 2009; 55: 750-61.

[273] Yang SJ, Park KY, Kim SH, *et al.* Antimicrobial resistance in *Salmonella* enterica serovars Enteritidis and Typhimurium isolated from animals in Korea: comparison of phenotypic and genotypic resistance characterization. Vet Microbiol 2002; 86: 295-301.

[274] Lysnyansky I, Mikula I, Gerchman I, Levisohn S. Rapid detection of a point mutation in the *parC* gene associated with decreased susceptibility to fluoroquinolones in *Mycoplasma bovis*. Antimicrob Agents Chemother 2009; 53: 4911-4.

[275] Rossi M, Hanninen ML, Revez J, Hannula M, Zanoni RG. Occurrence and species level diagnostics of *Campylobacter* spp., enteric *Helicobacter* spp. and *Anaerobiospirillum* spp. in healthy and diarrheic dogs and cats. Vet Microbiol 2008; 129: 304-14.

[276] Brisse S, Duijkeren E. Identification and antimicrobial susceptibility of 100 *Klebsiella* animal clinical isolates. Vet Microbiol 2005; 105: 307-12.

[277] Moreno A, Bello H, Guggiana D, Dominguez M, Gonzalez G. Extended-spectrum beta-lactamases belonging to CTX-M group produced by *Escherichia coli* strains isolated from companion animals treated with enrofloxacin. Vet Microbiol 2008; 129: 203-8.

[278] Vo AT, van Duijkeren E, Fluit AC, Gaastra W. A novel *Salmonella* genomic island 1 and rare integron types in *Salmonella* Typhimurium isolates from horses in The Netherlands. J Antimicrob Chemother 2007; 59: 594-9.

[279] Beckmann L, Muller M, Luber P, *et al.* [Suitability of SSCP-PCR analysis for molecular detection of quinolone resistance in *Campylobacter jejuni*]. Berl Munch Tierarztl Wochenschr 2003; 116: 487-90.

[280] Thibodeau A, Quessy S, Guevremont E, *et al.* Antibiotic resistance in *Escherichia coli* and *Enterococcus* spp. isolates from commercial broiler chickens receiving growth-promoting doses of bacitracin or virginiamycin. Can J Vet Res 2008; 72: 129-36.

CHAPTER 4

PCR-Based Diagnosis of Veterinary Bacterial Pathogens

Walter Lilenbaum[1*], Renata Fernandes Rabello[1] and Rubens Clayton da Silva Dias[2]

[1]Biomedical Institute, Fluminense Federal University, Niteroi, RJ, Brazil and [2]Biomedical Institute, Federal University of State of Rio de Janeiro, Rio de Janeiro, RJ, Brazil

Abstract: The ideal diagnostic method for veterinary purposes, particularly for the field practitioner, must be reliable, cost-effective and demonstrate good sensitivity and specificity. Although an ideal test with such characteristics does not yet exist, in a short horizon the most probable tests that could reach those goals are those based on molecular biology, particularly real-time PCR and its analogues. PCR-based methods, as a powerful tool for pathogen detection, have been frequently used in the identification of veterinary bacterial pathogens. This chapter focuses on the PCR method, some related important variations and their applications for the diagnosis of veterinary bacterial infections. The concepts such as restriction fragment length polymorphism analysis, multiplex PCR, nested PCR, allele-specific PCR, reverse transcription-PCR, real-time PCR and DNA sequencing are also discussed. This chapter particularly emphasizes the PCR-based diagnostic assays for *Brucella* sp., *Leptospira* sp., *Mycobacterium bovis*, *Staphylococcus aureus* and *Mycoplasma* sp. Real-time PCRs that could quantify the presence of the bacterial agents in a reliable way and identify the antimicrobial resistance and virulence factors genes have revolutionized veterinary medicine, making the diagnosis of infectious diseases rapid, reliable and cost-effective.

Keywords: PCR-based diagnosis; genotyping; multiplex PCR; nested PCR; allele-specific PCR; reverse transcription-PCR; Real-time PCR; *Brucella*; *Leptospira*; *Mycobacterium bovis*; *Staphylococcus aureus*; *Mycoplasma*.

1. INTRODUCTION

Molecular techniques are often used for the detection of uncultivable, slow-growing, fastidious or pleiotropic microorganisms. With high specificity and sensitivity, molecular techniques are also widely applied for the detection and characterization of common microorganisms. In addition, molecular methods have been used for strain typing that, together with classic epidemiology studies, have been crucial for the understanding of the transmission dynamics of infectious diseases [1, 2]. Genetic analysis of the rRNA gene and its spacer sequence is important for the bacterial taxonomy and phylogeny and bacterial identification [3-8]. Nevertheless, the analysis of rRNA sequences may have limitations, and the presence of several variable copies of ribosomal coding genes in microorganisms has been described [9]. Therefore, sequence analysis of several additional reference genes have also been used for identification, such as *rpoB* (RNA polymerase β-subunit), *gyrA* (DNA gyrase A subunit), *mdh* (malate dehydrogenase), *infB* (translation initiation factor 2), *phoE* (phosphoporine E) and *nifH* (nitrogenase reductase) [9-12]. For example, the comparison of the *rpoB* partial sequences of *Enterobacteriaceae* species revealed that divergence levels of sequences were clearly superior to those of 16S rRNA genes [11, 13]. Phylogenetic trees of *Enterobacteriaceae* isolates obtained with *rpoB* fit better with the classification of these organisms than 16S rRNA trees. Thus, *rpoB* sequencing has been frequently used as a means of identification and phylogenetic analysis of *Enterobacteriaceae*.

PCR-based methods using specific primers have been a powerful tool for the detection and identification of pathogenic microorganisms. The unquestionable success of detection assays based on the PCR has been largely due to its sensitivity, specificity and rapidity in comparison to many conventional diagnostic methods. For example, the time required for the detection and identification of *Brucella*, *Mycobacteria*,

***Address correspondence to Walter Lilenbaum:** Biomedical Institute, Fluminense Federal University, Niteroi, RJ, 24210-130, Brazil. Tel: +55.21.2629-2435 E-mail: mipwalt@vm.uff.br

Mycoplasma and other slow-growing and fastidious bacteria may be reduced from numerous days or even weeks in the classical methods, to a single day by PCR. While PCR methods show great advantages over the classical methods, several factors may affect the sensitivity and specificity of PCR. The complementarity of the primers to the target DNA and the annealing temperature of the PCR cycle are the most important factors for the production of specific PCR amplicons. Although *Taq* polymerase may tolerate some mismatches between the 3'end of the primer and the target DNA, it may resist non-complementarity at the 5'end of the primer. Therefore, numerous PCR variations have been designed to take advantage of these factors. The PCR data must be analyzed together with clinical significance; for example, repeated PCR positivity despite antibiotic treatment and/or serological non-reactivity and bacteriological sterility may mean chronicity or inadequacy of treatment.

Among molecular biological methods targeting nucleic acids, PCR has become the most popular diagnostic method in human and veterinary medicine, and food microbiology. Although numerous PCR assays have been developed for the detection and identification of bacterial pathogens, the majority of these assays are for research purposes. The further validation of these PCR assays is required before their commercial applications.

2. PCR METHODS

In the 1980s, a method for exponentially amplifying specific nucleic acid sequences using a DNA polymerase-mediated chain reaction was developed by Mullis and his colleagues and named polymerase chain reaction (PCR) [14]. This technique is able to amplify single copies of a DNA (or RNA) target into many copies. PCR has been a powerful tool, widely used in molecular analysis of nucleic acids. For diagnostic microbiology, PCR is widely used to amplify *in vitro* specific fragments of nucleic acids and is the basis for the detection of etiologic agents of infectious diseases and for the discrimination of non-pathogenic and pathogenic strains on the grounds of specific genes.

In PCR, a unique sequence of the target nucleic acid of interest is chosen for amplification. The specificity of the subsequent reaction is provided by two short oligonucleotides named primers. These oligonucleotides serve as primers for DNA polymerase-mediated DNA synthesis, using denatured target DNA as a template. PCR involves multiple cycles of a three-step process: denaturation (where the double-stranded DNA is separated into two single strands), annealing (when the primers hybridize to the DNA template), and extension (the primers are extended along the DNA template from the primer binding site). The success of PCR depends on temperature cycling ($92°C$ to $98°C$ for denaturation; about $40°C$ to $60°C$ for annealing; and $70°C$ to $75°C$ for primer extension). A DNA fragment located between the two annealed primers is multiplied. Virtually, a single copy of DNA may produce 2^{N+1} copies of the target after N cycles of amplification. Therefore, a single copy of DNA may be multiplied to more than one billion copies after 30 cycles in less than 1 hour. After amplification, the product may then be visualized *via* elctrophoresis on an agarose or acrylamide gel stained with ethidium bromide in conventional PCR, or quantified by specific fluorescence dye-labeled probes in real-time PCR.

Since only small modifications are necessary in a standard PCR protocol to achieve a desired goal, several PCR-based techniques have been developed to improve the accuracy and efficiency of PCR reactions. Indeed, in the past few years, a wide variety of methods have been developed to improve the specificity and sensitivity of PCR [15-30]. Thereby, the versatility of PCR has led to a large number of variants, such as the examples bellow.

2.1. PCR-Restriction Fragment Length Polymorphism Analysis (RFLP)

Some mutations generate or eliminate restriction enzyme recognition sites and may easily be detected by PCR-RFLP [31]. In this technique, the PCR products are digested by one or a combination of restriction enzymes and electrophoresed to detect polymorphisms or mutations which are seen as changes in DNA fragment sizes and patterns on the gel. When the rDNA is analyzed, the assay is referred to as PCR ribotyping.

2.2. Multiplex PCR

Multiplex PCR is a technique used for simultaneous amplification of several gene targets using multiple sets of PCR primer pairs in a single reaction [32, 33]. Therefore, multiplex PCR shows the advantage of detecting several pathogens simultaneously without the need of running several individual PCRs. This methodology, however, often shows reduced sensitivity due to the complex and unexpected interactions between different sets of primers, probes and nucleic acids, and due to the competition between PCR targets of high copy and low copy. Commonly, the multiplex PCR is named according to the number of primer pairs, such as duplex PCR (two pairs of primers) and triplex PCR (three pairs of primers).

2.3. Nested PCR

In nested PCR, a second pair of PCR primers, internal to the first one (nested), is used to amplify sequentially a single target in a second round of PCR. The product of the first PCR reaction is diluted and amplified with the second and nested primers. Thus, nested PCR is able to increase both the sensitivity and the specificity of PCR [34, 35].

2.4. Allele-Specific PCR

Usually PCR primers are designed from conserved regions of the genome and used to amplify a polymorphic area between them. However, in allele-specific PCR, the opposite is done. Since the 3'-end of the primer is essential in the extension of the primer, at least one of the primers should be designed to match/mismatch one of alleles at the 3'-end of the primer, generating a selective amplification of one of the alleles to detect single nucleotide polymorphism. A mismatched primer will not initiate amplification under stringent conditions; on the other hand, a matched primer will amplify the target. Therefore, this technique may conveniently determine the presence of an allele of a mutation or polymorphism in a target [36].

2.5. Reverse Transcription-PCR

Reverse transcription-PCR (RT-PCR) is the most sensitive and gold-standard technique for mRNA detection and quantitation [37]. Compared to RNAse protection assay and northern blot analysis, RT-PCR may be used to quantify mRNA levels from much smaller sample sizes. Indeed, this technique is sensitive enough to quantify mRNA from a single cell or pathogen. The traditional RT-PCR involves two steps: the RT reaction followed by PCR amplification. RNA is first reverse transcribed into cDNA using a reverse transcriptase. The resulting cDNA is used as a template for the subsequent PCR amplification. RT-PCR can also be carried out as a one-step RT-PCR in which all reaction components are mixed together in one tube prior to starting the reactions. One-step RT-PCR offers simplicity and convenience while reducing the possibility of contamination.

For quantifying mRNA, a competitive RT-PCR with internal standard RNAs should be applied. Competitive RT-PCR quantitates a message by comparing RT-PCR product signal intensity to a concentration curve generated by a synthetic competitor RNA sequence. The internal standard RNAs are added in a defined quantity to the RNA sample prior to the RT reaction. The resulting standard cDNA is co-amplified with the same primers as the endogenous target sequence. The competitor RNA transcript is designed for amplification by the same primers and with the same efficiency as the endogenous target. The competitor produces a different-sized product, approximately 50 nucleotides smaller, so that it can be distinguished from the endogenous target product by gel analysis. This method allows measurement of small differences in mRNA amount between RNA samples.

2.6. Real-Time PCR

A quantitative PCR (qPCR) method, named real-time PCR, has been developed and is becoming a valuable tool for diagnosing bacterial pathogens in clinical microbiology laboratories [38]. With regard to sensitivity, the real-time PCR is similar to conventional PCR. However, real-time PCR with the involvement of probes detects the increase in the amount of DNA as it is amplified and is therefore quantitative. There are distinct real-time PCR instruments that vary regarding to the nucleic acid probe formats supported, excitation and detection wavelengths, maximum number of samples per run and reaction volumes.

The use of SYBR Green in real-time PCR, one of the earliest formats and still a commonly employed dye, detects the accumulation of any double-stranded DNA product. The use of SYBR Green with instruments that perform a melting curve analysis permits detection of different amplification products based upon the G-C content and the length of the amplification product. Since this method provides sensitive detection but is not specific, SYBR Green assays are often used for screening assays where further analysis of specimens will confirm the results [39]. There are other ways of monitoring by using double strand intercalating dyes, such as SYTO9 and LC Green [39].

Sensitive and specific detection is possible with real-time PCR by using novel fluorescent probes. Three types of nucleic acid detection methods have been used most frequently with real-time PCR testing platforms in clinical microbiology: 5'nuclease (TaqMan probes), molecular beacons, and fluorescence resonance energy transfer (FRET) hybridization probes. All three of these real-time PCR instruments allow the use of all or some of the dyes used for TaqMan probes and molecular beacons. However, currently, only the LightCycler supports the FRET hybridization probe detection with melting curve analysis.

The significance of the name TaqMan probes comes from the videogame PacMan (Taq Polymerase + PacMan = TaqMan) because its mechanism is based on the PacMan principle. TaqMan probes are hydrolysis probes that increase the specificity of real-time PCR assays [40]. Its principle relies on the 5'-3' nuclease activity of *Taq* polymerase to cleave a dual-labelled probe during hybridization to the complementary target sequence and fluorophore-based detection. Therefore, as in other real-time PCR methods, the resulting fluorescence signal permits quantitative measurements of the accumulation of the product during the exponential stages of the PCR.

Similar to TaqMan probes, molecular beacon probes are oligonucleotide hybridization probes that have the two dye molecules attached to a single probe of oligonucleotide [41]. The first dye is a fluorescent dye, and the second can be either a quencher dye or another fluorescent dye which can absorb fluorescent light transferred from the first dye and re-emit light at a different wavelength. The proximity of the two dyes in the probe is determined by the hairpin shaped probe structure.

Contrasting with TaqMan and molecular beacon probes, FRET hybridization probes have the dyes attached separately to two probes designed to anneal next to each other in a head-to-tail configuration on target nucleic acid DNA [42]. The upstream probe has a fluorescent dye on the 3'end and the downstream probe has an acceptor dye on the 5' end. If both probes anneal to the target PCR product, fluorescence from the 3' dye excites the adjacent acceptor dye on the 5' end of the second probe. The second dye is excited and emits light at a third wavelength and this third wavelength is detected.

2.7. DNA Sequencing

DNA sequencing is an essential tool for validating the results of PCR and DNA hybridization-based assays. Described for the first time in 1977, the chain-terminator method is the basis for most DNA sequencing currently performed [43]. Then, radioisotope labeled dNTPs were used for DNA sequencing with the final sequence determined by manual inspection. This method was later replaced by automated DNA sequencing using fluorescent labeling (by using dye-primer or dye-terminator labeling) and it has become faster and more convenient. Currently, DNA sequencing is a common method of post amplification analysis of a PCR product. However, this technology usually requires user's expertise in software for edition and alignment of sequences, and for construction of phylogenetic trees. By analyzing regions of DNA conservation and variability, DNA sequencing becomes a powerful tool for the identification and differentiation of organisms [44-50].

3. TARGET GENES

PCR-based methods employ specific primers targeting a conserved gene within the genome. In fact, the development of a species-specific PCR was a significant step in the diagnosis of bacterial infections. The choice of the genomic region to be amplified will determine the specificity of detection from the onset. A characteristic target for the particular organism or a group of species should be selected, and the information provided by the nucleotide sequence is practically indispensable to assess the target's appropriateness.

The discovery of a large number of bacterial toxins and other virulence factors such as invasins and adhesins has led to a significantly better understanding of mechanisms of bacterial pathogenicity and powerful molecular methods for the accurate detection and identification of pathogenic bacteria. However, many pathogenic bacteria species do not have virulence factors. Therefore, the detection of bacteria and differentiation of pathogenic types from non- or less-pathogenic types must rely on other markers (such as 16S and 23S rRNA and DNA sequences of genes with unknown function) or still remains impossible. As described above, several protocols for molecular identification of pathogens were developed based on analysis of 16S rRNA genes and their spacer sequences. In addition, several housekeeping genes, such as *rpoB, gyrA, mdh, infB, phoE*, and *nifH*, have also been used for these purposes [9-13, 49].

The *rRNA* gene region has appeared to be the most prominent target in microbial detection. The popularity of this region is surely due to the highly conserved segments and moderately to highly variable segments that the region contains. Furthermore, the sequences of this gene are currently available for practically all microorganisms of human and veterinary health importance and are free to the public. Other gene targets for PCR detection, but not less important than rRNA, include insertion elements and repetitive sequences of some microorganisms such as IS*6110* [51] and direct repeat (DR) locus [52] of the mycobacterial chromosome. Such segments are present in multiple copies, which potentially increase the sensitivity of the detection method.

4. MOLECULAR STRAIN TYPING METHODS

Molecular typing of pathogens has become an important tool for the epidemiologic investigation of infectious diseases. Different molecular typing methods have enhanced our understanding of the epidemiology of several bacterial infectious diseases, and helped to determine modes of transmission of pathogens, as well as the source and the risk factors for infections.

Analysis of chromosomal DNA restriction patterns by pulsed field gel electrophoresis (PFGE) is an appropriate and discriminatory technique for typing different microorganisms [1]. However, PCR-based techniques such as random amplification of polymorphic DNA (RAPD), arbitrarily primed PCR (AP-PCR), enterobacterial repetitive intergenic consensus-PCR (ERIC-PCR), and repetitive extragenic palindromic-PCR (REP-PCR) analysis are faster and easier to perform [53-55], and provide useful discriminatory capabilities. PFGE and PCR-based methods have the disadvantage of relying on comparison of band patterns generated by gel electrophoresis, which is difficult to standardize. Comparability of patterns within and between laboratories requires strict adherence to established protocols. Even then, reproducibility is not always enough and normalization of patterns is complex. To overcome the poor portability problems of electrophoresis banding techniques, sequencing-based strain typing methods, such as multilocus sequence typing (MLST) [56, 57], have been developed. Unfortunately, the applicability of MLST to different pathogens in specific epidemiologic contexts is not well understood.

Although the numerous existing strain typing methods have been created primarily as a tool to determine the relatedness of pathogens, some of them (*e.g.,* RAPD-PCR, AP-PCR, ERIC-PCR, MLST) may be used for distinguishing bacterial species or individual strains, such as for *Bacillus* spp., *Brucella* spp., *Burkholderia pseudomallei, Escherichia coli, Listeria* spp. *Staphylococcus* spp., *Streptococcus pyogenes*, and *Streptomyces* spp. [56, 58, 59]. Other examples of methods developed for strain typing are the spoligotyping (a hybridization-PCR-based technique) and Low-stringency single specific primer PCR (LSSP-PCR) [60-63].

Spoligotyping is a method based on the polymorphism of the DR locus of the mycobacterial chromosome, developed as a genotyping tool to provide information on the structure of the DR region in individual *M. tuberculosis* strains and in different members of the *Mycobacterium tuberculosis* complex (MTbC) [52, 60-62, 64, 65]. This region is one of the hotspots for the IS*6110* integration and is present in all MTbC strains in a unique locus. In addition, the DR region is one of the most well studied loci of the MTbC genome showing considerable polymorphism among the strains. This locus contains multiple and well-conserved 36-bp repeats interspersed with non-repetitive short spacer sequences varying from 34 to 41-bp, which together are termed direct variable repeat (DVR) sequences. The order of the spacers was found to be well conserved in all isolates.

Currently, 94 different spacer sequences have been identified, and 43 of them are used for MTC strain differentiation. MTbC isolates vary in the numbers of 36-bp repeated elements and in the presence or absence of some spacers. As with other method mentioned above, though spoligotyping has been developed as a strain-typing tool, this technique is a useful method for the confirmation of specific identifications within MTbC. The strains are tested by hybridization between PCR-amplified DR regions to a membrane which consists of an array of covalently bound oligonucleotides representing the 43 different spacer sequences identified in the DR sequence of the *M. tuberculosis* H37Rv and *Mycobacterium bovis* BCG.

LSSP-PCR is an extremely simple PCR-based technique that permits detection of single or multiple mutations in gene-sized DNA fragments (at least up to 1 kb) [63]. Two PCR steps are necessary: in the first step, a DNA template is obtained by a specific PCR and purified from agarose gel; in the second one, this purified product is used as a template in the LSSP-PCR, which uses low-stringency conditions and only one primer, usually one of the two primers used in the first step. Thereby, the LSSP-PCR translates the underlying DNA sequence into a unique multiband "gene signature". Small changes, such as single-base mutations alter this signature significantly. PCR under these conditions results in a mixture of amplified DNA molecules, the composition of which is characteristic of the DNA segment assessed. Determination of the composition of the mixture can be achieved by subjecting the mixture to standard chromatographic or electrophoretic techniques to obtain a pattern of fragments which represents the "signature" of the sequence. LSSP-PCR presents almost unlimited molecular application where rapid and sensitive detection of mutation and sequence variations is important. Although it is essentially a strain typing technique, LSSP-PCR has also been used for identification purposes, such as detection of *Leptospira* directly from clinical specimens and discrimination of its serogroups from different animal reservoirs [63].

5. VETERINARY MEDICINE APPLICATIONS

PCR-based assays have been used for detection and characterization of bacteria extensively in the field of veterinary medicine. The main applications of these assays are for the diagnosis of specific agents of important bacterial infections of animals and for pathogens that are slow or difficult to grow and isolate in the clinical bacteriology laboratory. In addition, they are also important tools for epidemiological studies.

5.1. *Brucella* sp.

Brucella is a facultative intracellular pathogen that infects animals and human beings causing a disease known as brucellosis. Animal brucellosis, mainly in cows, is a major disease that causes abortion and infertility with severe economic losses. This disease also has a substantial impact on public health, since it is a foodborne and occupationally-acquired zoonosis [66].

At present, this genus consists of 10 recognized species based on phenotypic characteristics and host preference: *Brucella* (*B.*) *abortus* (cattle and bison), *B. melitensis* (goats and sheep), *B. suis* (pigs), *B. ovis* (sheep), *B. canis* (dogs), *B. neotomae* (wood rats) [67], and the recently describes species *B. ceti* (cetaceans), *B. pinnipedialis* (pinnipeds) [68], *B. microti* (vole *Microtus arvalis*) [69] and *B. inopinata* [70].

For the success of brucellosis control and eradication, reliable diagnostic methods must be employed. Serological tests, such as acidified antigen test (AAT) and many others are the major diagnostic tools for screening of anti-*Brucella* antibodies in the field. They are rapid, sensitive and easy to perform, but lack specificity due to cross-reactions with other bacteria. The 'gold standard' for the definitive diagnosis is the isolation and identification of the etiologic agent, especially in free areas where the positive predictive value of serological tests is very low. However, this method presents several drawbacks such as being time-consuming and complicated. Moreover, handling of this microorganism represents a potential hazard for laboratory personnel due to the zoonotic nature of most *Brucella* species. Furthermore, culture must be performed in well-equipped laboratories with highly skilled personnel [28, 66]. In order to overcome most of these limitations, numerous PCR-based assays have been developed for *Brucella* identification from cultures, animal/human tissues and animal products [23, 25, 28, 71, 72]. Some PCR-based assays were designed for identification at a genus level while others are able to differentiate among the species. Besides conventional assays, some authors have also developed real-time PCR-based assays [23, 25, 28, 71, 72].

The first PCR-based assays developed for *Brucella* identification were directed toward highly conserved loci in the genus. These assays tend to be simple and very robust, and can be utilized when differentiation of the species is not relevant such as in diagnosis of human brucellosis or contamination of food products [66]. For these applications, several targets have been employed such as the 43-kDa outer membrane protein gene, the 16S rRNA gene, the 31-kDa *Brucella* cell surface salt extractable protein (BCSP) gene, the perosamine synthetase gene and the 16S-23S intergenic spacer region [23, 28, 73-75]. Although these assays may be considered as specific, in some protocols cross-reactivity has been observed with the closely related genus *Ochrobactrum* [74, 75].

In spite of high degrees of genetic similarity among *Brucella* species and biovars, species-specific PCR-based assays have also been developed for the purpose of differentiation [71, 72, 76]. The differentiation among the species is mainly useful for epidemiological studies and the species-specific eradication programs.

Species-specific PCR-based assays have been designed mainly with highly specific primers and stringent assay conditions to reduce the risk of false-positive reactions. In this strategy, long primers are utilized in multiplex reactions. The assays have targeted the IS*711* (also known as IS*6501*), the *omp2* locus and the *rpoB* gene [23, 28, 77]. Moreover, genetic differences such as deletions, insertions and mutations are also used to design the primers [72, 76].

One of the most popular PCR-based assays for differentiation of *Brucella* species is the AMOS PCR which is based on the polymorphism arising from species-specific localization of the IS*711* in the *Brucella* chromosome [77]. IS*711* is a multi-copy insertion element that presents number and localization typically conserved in *Brucella* species. Initially, this assay could only differentiate among strains of *B. mellitensis*, *B. ovis* and some biovars of both *B. abortus* and *B. suis*. AMOS PCR was further modified to allow discrimination between the commonly used vaccine strains (S19 and RB51) and field strains, what is a critical point for any eradication program [78]. Recently, new modifications were made in order to allow the differentiation of some biovars previously not included [79].

Other multiplex PCR-based assays have been developed to detect and distinguish almost all presently recognized *Brucella* species, including certain biovars [72, 76]. PCR Bruce-Ladder was based on deletions, interruptions and mutations on the bacterial chromosome. Although this assay can also detect *Brucella* DNA from marine mammals, it is not suitable for discriminating between *B. ceti* and *B. pinnipedialis*. It can detect several *Brucella* biovars but does not discriminate them, and cannot detect *B. microti* as a separate species [76]. López-Goñi *et al.* (2008) [25] demonstrated the robustness and reproducibility of the Bruce-Ladder PCR by testing *Brucella* strains from five continents and from different hosts. In the multiplex PCR assay proposed by Huber *et al.* (2009) [72], all presently recognized *Brucella* species, except the most recent described species *B. inopita* were distinguished. A total of 19 primers are used in a single reaction: AMOS primers and primers based on additional specific loci of the IS*711* and other unique insertions and deletions. Nevertheless, its low sensitivity does not allow the detection of *Brucella* DNA directly from clinical samples without cultivation. Additionally, this multiplex PCR could not differentiate between *B. canis* and *B. suis* biovars [71, 72, 80]. Species-specific PCR-based assays have also been designed with semi-specific primers and moderately permissive assay conditions such as ERIC-PCR and REP-PCR [81, 82], and with random primers under very permissive conditions such as AP-PCR or RAPD-PCR [83, 84].

Most of the PCR-based assays for *Brucella* were developed for purified DNA obtained from culture. However, efforts have been made to identify the microorganisms directly from clinical specimens like aborted fetuses, associated maternal tissues, blood, milk, semen, nasal and vaginal secretions [85-93]. The detection of bacteria directly from clinical specimens is extremely useful for field diagnostic purposes.

5.2. *Leptospira* sp.

Leptospirosis is an important bacterial disease that affects humans and animals, including domestic and wild species. Recently, this disease has been considered an emerging infectious disease. In livestock animals, the main clinical signs are abortion or stillbirth in adults and a more severe form of the disease, frequently fatal, in calves causing high economic losses. Moreover, they are able to develop chronic renal infection and maintain persistent leptospiruria, disseminating bacteria to other animal species as well as to humans [94].

The taxonomy of leptospires has been continually evolving. Historically, there were only the species *Leptospira interrogans*, pathogenic to humans and animals, and *Leptospira biflexa*, nonpathogenic. *L. interrogans* has been subdivided into several serogroups and serovars. DNA hybridization techniques have promoted the identification of 17 genomospecies. The pathogenicity of a leptospiral serovar in a host animal varies depending on the host species and the geographic area [95].

The control of leptospirosis depends on rapid and accurate diagnosis. Serological, microbiological and molecular methods can be used to diagnose leptospirosis in animals. Serological methods, mainly microscopic agglutination test (MAT), are frequently used but show limitations such as cross-reactivity among serovars, presence of antibodies elicited by vaccination, interpretation difficulties, lack of antibodies in the early phase of the disease, and maintenance of a large number of *Leptospira* strains as sources of antigens. In addition, serology does not reliably identify the pathogen carriers, which is important in control programs [96, 97]. The microbiological method provides definitive diagnosis but it is not used for routine practice due to its poor sensitivity. Moreover, the leptospires are nutritionally stringent and grow slowly *in vitro* [95]. To overcome the limitations of serological and microbiological methods, PCR-based assays have been established for the rapid detection of *Leptospira* infections. PCR-based assays have lower detection limits than culture and may detect DNA from lysed or inactive microorganisms. PCR can be used to screen pathogens in herds and in semen for artificial reproduction [98, 20].

Different targets have been employed in PCR-based assays to detect leptospires including the 16S rRNA gene [24, 29, 99, 100], the 23S rRNA gene [22, 101], repetitive elements [102], endoflagellin genes [103], Lip32 gene [63], Hap 1 gene [104] and transcriptional regulator genes [105].

Several strategies have been developed for detection of leptospires such as conventional PCR [29, 100], nested PCR [20, 106], LSSP-PCR [63], real-time PCR [24] and PCR followed by RFLP [98]. Multiplex PCR for the molecular detection simultaneously of leptospires and other pathogenic bacteria have also been designed [20, 107].

Some targets and strategies allow detection and discrimination of the pathogenic and nonpathogenic leptospires [22, 24, 105, 106] and others allowed detection and grouping of the leptospires into serovars [98, 103] or serogroups [63]. Most studies have employed genus-specific PCR-based assays targeting the highly conserved gene among *Leptospira* species 16S rRNA [29, 100, 108]. Additionally, 16S rRNA gene-based primers have been designed to discriminate among pathogenic and non-pathogenic leptospires [24]. Moreover, PCR-based assays with primers based on 16S rRNA followed by RFLP can identify some serovars [98]. Primers designed to other targets have been used to detect pathogenic leptospires such as the 23S rRNA [22, 101], Lip32 [106] and HAP1 genes [99, 104]. The Lip32 and HAP1 genes, encoding another membrane lipoprotein and a hemolysis-associated protein-1, respectively, are present only in pathogenic leptospires and are conserved among them [63, 104]. Bomfim *et al.* (2006) [63] demonstrated the application of LSSP-PCR for the characterization of leptospires into serogroups directly from bovine urine samples by the analysis of polymorphisms present in a Lip32 genomic sequence fragment of *Leptospira*. The precise identification is important for epidemiological and public health surveillance. The applicability of genes encoding putative transcriptional regulators as potential targets for detection of pathogenic leptospires has also been investigated [105]. The established PCR tests can detect leptospires directly from clinical samples such as bovine and small ruminant urine [29, 63, 109], aborted bovine fetuses [107], small ruminant semen and vaginal fluids [100], bovine semen [98], canine semen [20], horse aqueous humor [110] and pig kidney tissues [24].

5.3. *Mycobacterium Bovis*

Mycobacterium (*M.*) *bovis*, a member of the *M. tuberculosis* complex (MTBC), is the causative agent of bovine tuberculosis (bTB). bTB is an infectious, chronic and progressive disease with a worldwide distribution. Besides the considerable economic implications, there are also public health implications since *M. bovis* is a zoonotic pathogen. For these reasons, bTB eradication programs have been implemented in a number of different countries [19, 111].

Most of the bTB eradication programs are based on screening of the herds using intradermal tuberculin tests ('skin tests') and removal of the reactors, coupled with slaughterhouse surveillance for undetected infection. The 'skin tests' are the international standard for *ante mortem* diagnosis of bTB in herds, but their results may be influenced by several factors such as infection stage and the severity of the disease as well as the presence of cross-reacting organisms, mainly environmental mycobacteria. Other tests such as IFN-γ assays, lymphocyte transformation and humoral immunity tests can be used to detect bTB in live animals. However, none of them permit a perfectly accurate determination of the *M. bovis* infection [112]. Due to the difficulty of obtaining specimens *in vivo,* bacteriological culturing is more commonly employed to confirm *M. bovis* infection in specimens from cattle slaughtered following a positive 'skin test' reaction. Although it is considered the 'gold standard' for the bTB diagnosis, culture has low sensitivity and, due to the fastidious growth of the bacteria, takes several weeks. An additional two or three weeks may be required for the identification of the isolate [113]. Considering the limitations of these conventional diagnostic methods, there has been an increasing interest in developing molecular methods for diagnosis of *M. bovis* infections.

Several PCR-based assays have been developed for detection of *M. bovis* directly from bovine clinical samples such as nasal secretions, milk, colostrum, blood, bronchoalveolar lavage and lymph nodes [17, 19, 86, 111, 113-116]. These assays offer advantages of sensitivity, flexibility and speed when compared to the bacterial culture. However, the sensitivity of these assays can be compromised by several factors such as the numbers of bacilli in the clinical specimens, DNA extraction method, pathological status of the animal, and the presence of PCR inhibitors in the samples [19, 62, 112].

The numbers of bacilli in clinical specimens may compromise the detection of mycobacterial DNA by PCR-based assays. The detection limit tends to vary according to the protocol employed. Mishra *et al.* (2005) [19] reported the detection of up to 50 fg of DNA (equivalent to 5 bacilli) by a nested PCR protocol and up to 0.1 ng (equivalent to 10^4 bacilli) by conventional PCR. A lower minimum detection limit, 10 fg of DNA (equivalent to 2 to 3 bacilli), was observed with a multiplex real-time PCR by Parra *et al.* (2008) [111].

Studies have demonstrated that different protocols to extract DNA from bovine tissues significantly affect the recovery rate of the nucleic acids [111, 113, 117]. The difficulties associated with DNA extraction are attributed to low number of bacteria and the strong fibrosis and calcification of affected tissues that hamper the release of DNA [118]. A number of DNA extraction procedures have been described [17, 19, 111, 113, 119, 120]. The nucleic acid sequence capture procedure has improved the DNA extraction efficiency and has enabled the detection of mycobacteria in paucibacillary forms of TB [17]. Silica-based methods of DNA extraction have been widely evaluated and found to be one of the most efficient [113].

Sensitivity of PCR-based assays, as well as of other diagnostic methods, varies according to the pathological state of the animal [111, 121]. The sensitivity is higher when the animals tested have widespread lesions compatible with bTB. Parra *et al.* (2008) [111] observed sensitivity of 58.3%, 75% and 80.6% for tested animals without visible lesions, with initial lesions and with generalized lesions, respectively. The sensitivity was lower in relation to the culture only when animals with no lesions visible were tested.

The primers have been designed to detect species of MTBC (MTBC-specific targets) or to detect and differentiate species of MTBC (species-specific targets). MTBC-specific targets include multiple IS (*e.g.,* IS*6110,* IS*1081*) [113, 122], *hsp65* [86], 16S rRNA and 23S DNA genes [18]. Multiple IS are potential targets for PCR design to detect paucibacillary infections, improving the detection limit. Initially, IS*6110* was considered a potential target to differentiate MTBC from other mycobacteria. However, further studies have shown that IS*6110* exists in some non-MTBC species while some MTBC strains lack IS*6110* [18]. The gene *hsp65,* encoding a putative host cell receptor binding protein, is well conserved within the MTBC [86].

Bovides can also be infected with *M. tuberculosis* by exposure to infected humans that are shedding the microorganisms. Furthermore, the speciation of the members of the MTBC would help to determine the origin of infection and formulate appropriate public health strategies to interrupt the transmission route [19]. The targets used to detect and differentiate the species of the MTBC include *hupB* [19], *pncA* [123], *gyrB* [124], *oxyR* [125] and *katG* genes [126]. The gene *hupB* encodes a histone-like protein that is present in the *M. bovis* and *M. tuberculosis* genome, and there is a 27-bp difference in the C-terminal parts between *M. bovis* and *M. tuberculosis*. Different PCR-based methods such as PCR-RFLP and nested-PCR have targeted this gene [19]. The gene *pncA*, encoding the pyrazinamidase, contains a conserved point mutation in *M. bovis*. The polymorphism of this gene allows differentiation between *M. bovis* and *M. tuberculosis* through PCR-RFLP assays [123]. Single nucleotide changes in the *gyrB* [124], *oxyR* [127] and *katG* [126] genes have also been exploited for the differentiation of these two species. The *mtp40* gene, which was previously considered specific for *M. tuberculosis* [128, 129], was used to distinguish between *M. tuberculosis* and *M. bovis*. However, the use of this gene for species differentiation has recently been invalidated, because the gene is not present in all *M. tuberculosis* strains or absent in all *M. bovis* strains [127, 130]. The region of difference 4 (RD4) is present in the genome of all members of the MTBC except *M. bovis*. Some protocols use a set of primers designed to this region combined with another set that amplify a region present in all mycobacteria [113, 131].

Different PCR-based methods have been evaluated to detect and distinguish MTBC, including nested PCR [19, 115], spoligotyping [62], multiplex PCR [131], PCR-RFLP [123], real-time PCR [17, 111] and RT-PCR [114]. Recently, a blood-based real-time RT-PCR assay for quantification of IFN-γ mRNA was developed to diagnose tuberculosis in domestic (such as cattle, sheep, goat) and wildlife species (such as reindeer, white-tailed deer, red deer, elk, and bison). Detection of mRNA is an alternative method of demonstrating IFN-γ induction by mycobacteria [114]. In some countries, a complete control of bTB has been hampered by the existence of wildlife reservoirs of *M. bovis*. 'Skin tests' are impractical for use with many of these species and IFN-γ-based assays are unavailable because of a lack of species-specific immunological reagents. The real-time RT-PCR assay is simple, rapid and sensitive. Moreover, it does not compromise the immune status of the animal, allowing for repeat testing if necessary. The IFN-γ m-RNA responses correlated well with IFN-γ cytokine production and showed performance superior to that either lymphocyte proliferation or IFN-γ cytokine enzyme-linked immunosorbent assay methods. This assay would be particularly useful for wildlife species because it would require only a single handling event. The specificity of the test is impacted by cross-reactive sensitization to *M. bovis* antigens induced by exposure to *Mycobacterium avium* and other nontuberculous *Mycobacterium*sp. [114].

5.4. *Mycoplasma* sp.

Mycoplasma species are etiologic agents of serious and diverse diseases of livestock, and result in great economic losses. Some of these diseases are classified as being notifiable diseases by the World Organization for Animal Health (OIE) [132].

Mycoplasma mycoides subsp. *mycoides* small colony variant (MmmSC) and *Mycoplasma capricolum* subsp. *capripneumoniae* (Mccp) can cause severe respiratory infections in cattle, named contagious bovine pleuropneumonia, and a similar disease in goats, named contagious caprine pleuropneumonia [133]. *Mycoplasma agalactiae* is the main etiological agent of contagious agalactia in small ruminants while other mycoplasmas may also be involved with agalactia such as *M. capricolum* subsp. *capricolum* (Mcc), *Mycoplasma mycoides* subsp. *capri* (Mmc) and *M. mycoides* subsp. *mycoides* large colony variant (MmmLC) [133, 134]. *M. bovis* is a major cause of pneumonia and arthritis in calves, as well as mastitis and genital infections in adult cows [134]. *M. Hyopneumoniae* results in enzootic pneumonia (EP), a chronic respiratory disease that affects mainly finishing pigs [135], while *Mycoplasma gallisepticum* and *Mycoplasma synoviae* can cause chronic respiratory disease and infectious synovitis respectively, in poultry [136].

Since *Mycoplasma* infections do not present characteristic clinical signs, laboratory diagnosis is required. Diagnosis can be achieved through bacterial culture, serologic tests, post-mortem examination or PCR-based assays. Although bacterial culture is the 'gold standard' for diagnosis of mycoplasma infections, it is laborious and time-consuming since mycoplasmas require several weeks to grow. In addition, pathogenic avian mycoplasma may be impaired to grow by saprophytic mycoplasmas present in the upper respiratory tract [136]. Serologic tests have been frequently used to diagnose mycoplasma infections, mainly avian mycoplasma infections and *M. hyopneumoniae* infections in swine. However, serologic tests are of lesser value for detection of early infections and subclinical infections [137, 138], and are unable to discriminate natural infection from vaccination in swine [135]. Post-mortem inspections in abattoir surveillance or field necropsy are frequently used to diagnose *M. hyopneumoniae* infection, but enzootic pneumoniae lesions are not pathognomonic [135]. To overcome the limitations of these diagnosis techniques, PCR-based assays have gained a pivotal role in the diagnosis of mycoplasma infections [134, 139].

5.4.1. Mycoplasma Mycoides Cluster

Some of the *Mycoplasma* species that are pathogenic for ruminants are closely related and constitute the *Mycoplasma mycoides* cluster, which is formed by MmmSC, MmmLC, Mmc, Mcc, Mccp and *Mycoplasma* sp. bovine group 7 of Leach (MBG7) [140]. The high degree of similarity among them consequently generates difficulties in diagnosis. In addition, high intra-species variability is observed among some of these species such as MmmLC/Mmc and Mcc [141].

The introduction of PCR-based assays has made the diagnosis much more sensitive and reliable [26]. These assays have been designed to detect and identify the *M. mycoides* cluster or some of the species of this cluster as MmmSC [142], MmmLC [143], Mcc and Mccp [144]. However, some protocols for identification at the species level have shown to be unreliable for field isolates [145]. Besides classical PCR-based assays, some real-time PCR protocols have also been developed [26, 146]. The targets employed include CAP-21 gene [147], 16S rRNA gene [148], lipoprotein genes [132, 143, 149], IS [150, 151], membrane protein genes [132, 141], glucokinase gene [142], and others [144]. Some of these assays require further analytical steps using restriction enzyme digestion or DNA sequencing.

5.4.2. M. agalactiae and M. bovis

The routine analysis method used by diagnostic laboratories for identifying *M. agalactiae* in clinical samples or milk tanks is based on microbiological culture [30]. However, differentiation between *M. agalactiae* and *M. bovis* is problematic when only serological and biochemical tests are used. Both agents share a considerable number of related proteins and common epitopes [134], and it is problematic to differentiate between *M. agalactiae* and *M. bovis* by serological and biochemical tests.

Due to the limitations of these methods, PCR-based methods for directly detecting *M. agalactiae* in milk or other fluids have been established. In the countries where sheep and goat breeding is important, a compulsory diagnosis scheme is implemented to assess the presence of *M. agalactiae* in milk bulks or in the flocks by PCR assay [152]. Techniques like PCR have also been used to detect ill animals and latent carriers among males for insemination centers. Healthy male goats can carry mycoplasmas without exhibiting a serological response, and this limits the practical use of serological tests [153].

Several PCR-based methods that directly detect and identify *M. agalactiae* in milk bulks or in animal samples have been described [30]. Some of these methods are based on amplification of the 16S rRNA gene [134, 154-156]. However, 16S rRNA gene sequences in *M. agalactiae* and *M. bovis* share 99.8% similarity, affecting the specificity of methods based on the amplification of this gene [30]. Therefore, efforts have been made to identify other targets that allow the detection and differentiation of these species. Subramaniam *et al.* (1998) [157] developed a PCR-RFLP assay targeted at the DNA repair *uvrC* gene, which was shown to clearly differentiate between *M. bovis* and *M. agalactiae*. Modifications of this assay were made by Thomas *et al.* (2004) [139], aiming to reduce the costs and make the test more suitable for routine diagnostics. The *p40* gene, encoding an adhesin that plays a key role in cytoadhesion of *M. agalactiae,* has also been proposed as a good target for real-time PCR assay [30]. The *mb-mp81* gene encodes a membrane lipoprotein P81, present in the both *M. agalactiae* and *M. bovis* genome [158-160]. Some PCR-based strategies are PCR-RFLP [158] and other real-time PCRs [159, 160].

Several other PCR-based assays have been described for the detection of *M. bovis* [161-163]. The *oppD/F* gene region of *M. bovis*, which encodes ATP-binding proteins of the ABC-transporter family, was also used as a specific target region in a PCR test for direct detection of the organism in milk [147, 164]. Bashiruddin *et al.* (2005) [134] evaluated the performance of different PCR systems for the identification and differentiation of *M. agalactiae* and *M. bovis*. The specificity of the different detection systems appears to be high. However, assays targeting housekeeping genes (*oppD/F* and *uvrC*) have shown to be very reliable compared to assays targeting rRNA genes.

5.4.3. M. hyopneumoniae

Several PCR techniques have also been developed for *M. hyopneumoniae* DNA detection in samples such as bronchoalveolar lavage fluid (BALF), lung tissues, nasal swabs and tracheo-bronchial lavage [165-167]. For the detection of *M. hyopneumoniae*, PCR methods are more rapid than bacteriological culture and are relatively inexpensive. The best samples for PCR detection of *M. hyopneumoniae* are tracheo-bronchial swabs or BALF because *M. Hyopneumoniae* attaches to the ciliated epithelium of the airways. The detection of *M. hyopneumoniae* by PCR provides a more precise method of determining the infectious status of the animals, better than the seroconversion approach by serological methods [135].

5.4.4. Avian Mycoplasmas

PCR-based tests are now routinely used for detecting pathogenic avian mycoplasmas. DNA targets for PCR can be extracted from cultures or directly from swabs [168]. PCR-based assays are based on the 16S rRNA gene [169, 170], 16S-23S rRNA intergenic spacer region [136, 171] and *vlhA* haemagglutinin (HA) gene [168, 172-174]. The 16S-23S rDNA intergenic spacer regions (ISRs) of most avian mycoplasmas are relatively small, species-specific, and intra-specific conserved and can serve as an ideal genomic target for PCR diagnosis [136]. PCR targeting the *vlhA* gene can be also used to type strains of *M. synoviae* due to sequence variability in the N-terminal region of the gene among strains [168, 172, 175]. Raviv *et al.* (2009) [136] described the development of a complete array of species-specific real-time TaqMan PCR assays for the four pathogenic avian mycoplasmas, including *M. gallisepticum* and *M. synoviae*, and have proposed the incorporation of these assays into routine diagnostics.

5.5. Staphylococcus Aureus and Other Coagulase-Positive Staphyolococci (CoPS)

The genus *Staphylococcus* is constituted by several species, and some of them are important pathogens in humans and animals. *Staphylococcus* (*S.*) *aureus* is one of the most frequently isolated species causing a range of pathologies. In animals, the main illness caused by this species is mastitis in ruminants. However, other staphylococcal species have emerged as etiologic agents of this disease such as the coagulase-negative staphylococci species [27, 176, 177].

Mastitis is one of the most important diseases that occur in dairy cattle, being responsible for huge economic losses. Therefore, the rapid and early diagnosis of mastitis is imperative for its control and maintaining healthy cattle. Currently, diagnostic assays for mastitis include measurement of somatic cell counts (SCCs), enzymatic analysis, California Milk Test and culture of the milk. The culture of milk is still considered the 'gold standard' for the detection and identification of mastitis-causing microorganisms. Besides identifying the causative bacteria, it allows for the determination of their sensitivity to antibiotics. However, species identification is time-consuming and negative milk samples represent a large proportion of samples analyzed [178].

S. aureus is the leading cause of intramammary infections in ruminants. Culture has the sensitivity ranges of 41 to 100% for diagnosis of *S. aureus* intramammary infections [179]. The main reasons for low sensitivity include intervals of low levels of *S. aureus*s shedding in the milk, intracellular localization of bacteria into the host cells and antimicrobial residues [177].

PCR-based methods have been investigated as great tools for rapid and sensitive detection of *S. aureus*. These approaches detected *S. aureus* in a milk sample which did not show bacterial growth in conventional culturing [178]. Besides the PCR-based assays that have been developed to detect *S. aureus* [21], multiplex and real-time PCR have been designed to detect simultaneously different mastitis-causing organisms in milk samples [178, 180]. Commercial PCR-based mastitis tests are also available and are capable of detecting several of the major mastitis-associated pathogens, including *Escherichia coli*, *S. aureus*, coagulase-negative *Staphylococcus*, *Streptococcus agalactiae* and *Streptococcus uberis* [178].

Several molecular targets have been exploited for the molecular identification of *Staphylococcus* species, including the 16S to 23S rRNA spacer region [181], 16S rRNA and 23S rRNA genes [180, 182], *nuc* [21], *femA* [183], *sodA* [184], *rpoB* [185] and *groEL* genes [27]. The *groEL* gene, encoding heat shock protein, was proven to be an ideal universal DNA target for species identification because it presents a well-conserved DNA sequence within a given species, but with sufficient sequence variations to allow species-specific identification. PCR-RFLP assay targeting this gene has been successfully shown to be a reliable tool used for the identification of the main *Staphylococcus* species involved in bovine mastitis [27].

Real time PCR assays have been also developed to direct quantification of *S. aureus* in dairy products in countries that have regulations to classify these products. Traditional microbiological methods such as the plate count method or the most-probable-number (MPN) technique for quantification of *S. aureus* in milk are time-consuming and require anywhere from two, up to six days for quantification and detection [186].

In small animals, *Staphylococcus intermedius* was considered responsible for most cases of the canine pyoderma, a major skin disease of dogs. However, molecular analysis has shown that the causative agents consist of three distinct species, including *S. intermedius*, *Staphylococcus pseudintermedius* and *Staphylococcus delphini* (known as *S. intermedius* group or SIG). Of these species, the common etiologic agent of canine pyoderma seems to be *S. pseudintermedius*. Phenotypic methods are unable to discriminate *S. pseudintermedius* from *S. delphini*, and DNA sequencing may be required for accurate identification. Bannoehr *et al.* (2009) [187] have developed a rapid, simple and robust PCR-RFLP approach for clinical identification of *S. pseudintermedius*, in which a fragment of *pta* gene is amplified and digested with a restriction enzyme.

6. FUTURE PERSPECTIVES

Real-time PCRs have essentially been established for all types of reported bacterial pathogens, such as *Borrelia*, *Brucella*, *Helicobacter*, *Leptospira*, *Mycobacterium* and *Mycoplasma*, among many others. Those tests have in common the fact that they can detect PCR products during the amplification steps, and do not require the further steps of opening PCR tubes and running the gel electrophoresis, renedering these those tests faster and more reliable than the traditional PCR protocols. Although the real-time PCRs have not been adequately tested in controlled field trials for some infectious agents, the possibility of obtaining a reliable direct diagnosis under field conditions at an affordable price makes these tests the most promising trends for the future of veterinary diagnostics [188].

Another very promising line of research is based on multiplex PCRs. Although some of them have already been described for various bacteria, the possibility of concentrating the diagnosis of the main infections that occur in a livestock species or in a particular region in only one test is very interesting. Besides detecting the infectious agent, PCR-based tests may also be used for other diverse and important applications, such as detection of virulence factors and detections of antimicrobial resistance genes. In some bacteria, such as *Staphylococcus* sp., the detection of the agent per se is not sufficient for good therapeutics. The sensitivity of that strain to the antimicrobial must be known, since it can vary enormously among the different strains. Therefore, the possibility of simultaneously detecting the agent and the presence of the antimicrobial resistance genes, or at least the frequency of them, may have a great impact on veterinary medicine. Many of the resistance genes can already be detected separately in those bacteria, such as the *mecA* gene that encodes for the methicilin resistance and others. A multiplex PCR that could combine not only the detection of the bacteria but also some of the most frequent and important resistance genes would enormously simplify the diagnosis of the disease determined by those agents and improve therapeutics and control measures.

By analogy, the detection of virulence factors in some pathogens may also be very important. Many agents are members of the normal microbiota, and their presence in clinical specimens may not necessarily be considered as an unquestionable indication that the bacterium is the etiological agent of a disease. That point has been described by some authors as one of the disadvantages of the PCR-based methods on clinical specimens of veterinary origin. Nevertheless, if a multiplex PCR could detect not only the presence of the bacterial DNA but also some of the main virulence genes, which have already been described for the majority of infectious agents, then the effective diagnosis of the pathogen involved in the clinical disease could be demonstrated.

In conclusion, the molecular tests, such as real-time PCR, that could quantify the presence of the bacterial agents and demonstrate the antimicrobial resistance and virulence factors genes, would have a huge and revolutionary impact on veterinary medicine, making the diagnosis of infectious diseases faster, cost-effective and reliable.

REFERENCES

[1]   Tenover FC, Arbeit RD, Goering RV, *et al.* Interpreting chromosomal DNA restriction patterns produced by pulsed-field gel electrophoresis: criteria for bacterial strain typing. J Clin Microbiol 1995; 33: 2233-39.

[2]   Tenover FC, Arbei, RD, Goering RV. How to select and interpret molecular strain typing methods for epidemiological studies of bacterial infections: a review for healthcare epidemiologists. Molecular Typing Working Group of the Society for Healthcare Epidemiology of America. Infect Control Hosp Epidemiol 1997; 18: 426-39.

[3]   Van de Peer Y, Chapelle S, De Watcher R. A quantitative map of nucleotide substitution rates in bacterial rRNA. Nucleic Acids Res 1996; 24: 3381-91.

[4]   Clarridge JE III. Impact of 16S rRNA gene sequence analysis for identification of bacteria on clinical microbiology and infectious diseases. Clin Microbiol Rev 2004; 17: 840-62, 2004.

[5]   Hoshino T, Furukawa K, Tsuneda S, Inamori Y. RNA microarray for estimating relative abundance of 16S rRNA in microbial communities. J Microbiol Methods 2007; 69: 406-10.

[6]   Muir P, Oldenhoff WE, Hudson AP, *et al.* Detection of DNA from a range of bacterial species in the knee joints of dogs with inflammatory knee arthritis and associated degenerative anterior cruciate ligament rupture. Microb Pathog 2007; 42: 47-55.

[7]   Ni YQ, Yang Y, Bao JT, *et al.* Inter- and intraspecific genomic variability of the 16S-23S intergenic spacer regions (ISR) in representatives of Acidithiobacillus thiooxidans and Acidithiobacillus ferrooxidans. FEMS Microbiol Lett 2007; 270: 58-66.

[8]   Ruppitsch W, Stoger A, Indra A, *et al.* Suitability of partial 16S ribosomal RNA gene sequence analysis for the identification of dangerous bacterial pathogens. J Appl Microbiol 2007; 102: 852-59.

[9]   Rosenblueth, M, Martínez L, Silva J, Martínez-Romero E. *Klebsiella variicola*, a novel species with clinical and plant-associated isolates. Syst Appl Microbiol 2004; 27: 27-35.

[10]   Carter JS, Bowden FJ, Bastian I, *et al.* Phylogenetic evidence for reclassification of Calymmatobacteriumgranulomatis as *Klebsiella granulomatis comb*. nov. Int J Syst Bacteriol 1999; 49: 1695-700.

[11]   Drancourt M, Bollet C, Carta A, Rousselier P. Phylogenetic analyses of Klebsiella species delineate Klebsiella and *Raoultella gen. nov.*, with description of Raoultella ornithinolytica comb. nov., Raoultella terrigena comb. nov. and Raoultella planticola comb. nov. Int J Sys Evol Microbiol 2001; 51: 925-32.

[12]   Li X, Zhang D, Chen F, *et al. Klebsiella singaporensis* sp. nov., a novel isomaltulose-producing bacterium. Int J Syst Evol Microbiol 2004; 54: 2131-36.

[13]   Mollet C, Drancourt M, Raoult D. rpoB sequence analysis as a novel bases for bacterial identification. Mol Microbiol 1997; 26: 1005-11.

[14]   Mullis K, Faloona F, Scharf S, *et al.* Specific enzymatic amplification of DNA *in vitro*: the polymerase chain reaction. Biotechnology 1992; 24: 17-27.

[15]   Hill W E, Keasler S P. Identification of foodborne pathogens by nucleic acid hybridization. Int J Food Microbiol 1991; 12: 67-76.

[16]   D'Aoust J-Y. Salmonella and the international food trade. Int J Food Microbiol 1994; 24: 11-31.

[17]   Taylor MJ, Hughes MS, Skuce RA, Neill SD. Detection of *Mycobacterium bovis* in bovine clinical specimens using real-time fluorescence and fluorescence resonance energy transfer probe rapid-cycle PCR. J Clin Microbiol 2001; 39: 1272-1278.

[18]   Kurabachew M, Enger Ø, Sandaa RA, *et al.* A multiplex polymerase chain reaction assay for genus-, group- and species-specific detection of mycobacteria. Diagn Microbiol Infect Dis 2004; 49: 99-104.

[19]   Mishra A, Singhal A, Chauhan DS, *et al.* Direct detection and identification of Mycobacterium tuberculosis and Mycobacterium bovis in bovine samples by a novel nested PCR assay: correlation with conventional techniques. J Clin Microbiol 2005; 43: 5670-78.

[20]   Kim S, Lee DS, Suzuki H, Watari M. Detection of Brucella canis and Leptospira interrogans in canine semen by multiplex nested PCR. J Vet Med Sci 2006; 68: 615-18.

[21]   Graber HU, Casey MG, Naskova J, *et al.* Development o ahighly sensitive and specific assay to detect Staphylococcus aureus in bovine mastitic milk. J Dairy Sci 2007; 90: 4661-69.

[22]   Kositanont U, Rugsasuk S, Leelaporn A, *et al.* Detection and differentiation between pathogenic and saprophytic Leptospira sp. by multiplex polymerase chain reaction. Diagn Microbiol Infect Dis 2007; 57: 117-22.

[23]   Mukherjee F, Jain J, Patel V, Nair M. Multiple genus-specific markers in PCR assays improve the specificity and sensitivity of diagnosis of brucellosis in field animals. J Med Microbiol 2007; 56: 1309-16.

[24]   Fearnley C, Wakeley PR, Gallego-Beltran J, Dalley C *et al.* The development of a real-time PCR to detect pathogenic Leptospira species in kidney tissue. Res Vet Sci 2008; 85: 8-16.

[25]   López-Goñi I, García-Yoldi D, Marín CM, *et al.* Evaluation of a multiplex PCR assay (Bruce-ladder) for molecular typing of all Brucella species, including the vaccine strains. J Clin Microbiol 2008; 46: 3484-87.

[26]   Lorenzon S, Manso-Silva L, Thiaucourt F. Specific real-time PCR assays for the detection and quantification of Mycoplasma mycoides subsp. mycoides SC and Mycoplasma capricolum subsp. capripneumoniae. Mol Cell Probes 2008; 22: 324-28.

[27]   Santos, OCS, Barros EM, Brito MAVP, *et al.* Identification of coagulase-negative staphylococci from bovine mastitis using RFLP-PCR of the groEL gene. Vet Microbiol 2008; 130: 134-40.

[28]   Bounaadja L, Albert D, Chénais B, *et al.* Real-time PCR for identification of Brucella sp. : a comparative study of IS711, bcsp31 and per target genes. Vet Microbiol 2009; 137: 156-64.

[29]   Lilenbaum W, Varges R, Ristow P, *et al.* Identification of Leptospirasp. carriers among seroreactive goats and sheep by polymerase chain reaction. Res Vet Sci 2009; 87: 16-19.

[30]   Oravcová K, López-Enríquez L, Rodríguez-Lázaro D, Hernández M. Mycoplasma agalactiaep40 gene, a novel marker for diagnosis of contagious agalactia in sheep by real-time PCR: assessment of analytical performance and in-house validation using naturally contaminated milk samples. J Clin Microbiol 2009; 47: 445-50.

[31]   Haliassos A, Chomel JC, Tesson L, *et al.* Modification of enzymatically amplified DNA for the detection of point mutations. Nucleic Acids Res 1989; 17: 3606.

[32]   Chamberlain JS, Gibbs RA, Rainier JL, Caskey CT. Multiplex PCR for the diagnosis of Duchenne muscular dystrophy. In: Innis MA, Gelfand DH, Sninsky JJ, White TJ, Eds. PCR Protocols: A Guide to Methods and Applications. New York, Academic Press, 1990; pp. 272-81.

[33]   Frankel G, Giron JA, Valmassoi J, Schoolnik GK. Multi-gene amplifications: simultaneous detection of three virulence genes in diarrhoeal stool. Mol Microbiol 1989; 3: 1729-34.

[34]   Erlich HA, Gelfand D, Sninsky JJ. Recent advances in the polymerase chain reaction. Science 1991; 252: 1643-51.

[35]   Haqqi TM, Sarkar G, David CS, Sommer SS. Specific amplification of a refractory segment of genomic DNA. Nucleic Acids Res 1988; 16: 11844.

[36]   Ferrie RM, Schwarz MJ, Robertson NH, *et al.* Development, multiplexing, and application of ARMS tests for common mutations in CFTR gene. Am J Hum Genet 1992; 51: 251-62.

[37]   Myers TW, Gelfand DH. Reverse transcription and DNA amplification by a Thermus thermophilus DNA polymerase. Biochemistry 1991; 30: 7661-66.

[38]   Espy MJ, Uhl JR, Sloan LM, *et al.* Real-time PCR in clinical microbiology: applications for routine laboratory testing. Clin Microbiol Rev 2006; 19: 165-256.

[39]   Monis PT, Giglio S, Saint CP. Comparison of SYTO9 and SYBR Green I for real-time polymerase chain reaction and investigation of the effect of dye concentration on amplifcation and DNA melting curve analysis. Anal Biochem 2005; 340: 24-34.

[40]   Heid CA, Stevens J, Livak KJ, Williams PM. Real time quantitative PCR. Gen Res 1996; 6: 986-94.

[41]   Piatek AS, Tyagi S, Pol AC, *et al.* Molecular beacon sequence analysis for detecting drug resistance in Mycobacterium tuberculosis. Nat Biotechnol 1998; 16: 359-63.

[42]   Chen X, Kwok PY. Homogeneous genotyping assays for single nucleotide polymorphisms with fluorescence resonance energy transfer detection. Genet Anal 1999; 14: 157-63.

[43]   Sanger F, Nicklen S, Coulson AR. DNA sequencing with chain terminating inhibitors. Proc Natl Acad Sci USA 1977; 74: 5463-67.

[44]   Cursons RTM, Jeyerajah E, Sleigh JW. The use of polymerase chain reaction to detect septicemia in critically ill patients. Crit Care Med 1999; 27: 937-40.

[45]   Rantakokko-Jalava K, Nikkari S, Jalava J, *et al.* Direct amplification of rRNA genes in diagnosis of bacterial infections. J Clin Microbiol 2000; 38: 32-39.

[46]   Sleigh J, Cursons R, Pine ML. Detection of bacteremia in critically ill patients using 16S rDNA polymerase chain reaction and DNA sequencing. Intensive Care Med 2001; 27: 1269-73.

[47]   Misawa N, Kawashima K, Kondo F, *et al.* Isolation and characterization of Campylobacter, Helicobacter, and Anaerobiospirillum strains from a puppy with bloody diarrhea. Veter Microbiol 2002; 87: 353-64.

[48]   Schuurman T, de Boer, RF, Kooistra-Smid AMD, van Zwet AAl. Prospective study of use of PCR amplification and sequencing of 16S ribosomal DNA from cerebrospinal fluid for diagnosis of bacterial meningitides in a clinical setting. J Clin Microbiol 2004; 42: 734-40.

[49]   Alves MS, Dias RCS, Castro ACD, *et al.* Identification of clinical isolates of indole-positive and indole-negative Klebsiellasp. J Clin Microbiol 2006; 44: 3640-46.

[50]   Ratanart R, Cazzavillan S, Ricci Z, *et al.* Usefulness of a molecular strategy for the detection of bacterial DNA in patients with severe sepsis undergoing continuous renal replacement therapy. Blood Purif 2007; 25: 106-11.

[51]   Thierry D, Cave MD, Eisenach KD, *et al.* IS6110, an IS-like element of Mycobacterium tuberculosis complex. Nucleic Acids Res 1990; 18: 188.

[52]   Kamerbeek J, Schouls L, Kolk A, *et al.* Simultaneous detection and strain differentiation of Mycobacterium tuberculosis for diagnosis and epidemiology. J Clin Microbiol 1997; 35: 907-14.

[53]   Gori A, Espinasse F, Deplano A, *et al.* Comparison of pulsed-field gel electrophoresis and randomly amplified DNA polymorphism analysis for typing extended-spectrum-β-lactamase-producing Klebsiella pneumoniae. J Clin Microbiol 1996; 34: 2448-53.

[54]   Gazouli M, Kaufmann ME, Tzelepi E, *et al.* Study of an outbreak of cefoxitin-resistant Klebsiella pneumoniae in a general hospital. J Clin Microbiol 1997; 35: 508-10.

[55]   Cartelle M, Tomas MdM, Pertega S, *et al.* Risk factors for colonization and infection in a hospital outbreak caused by a strain of Klebsiella pneumoniae with reduced susceptibility to expanded-spectrum cephalosporins, J. Clin. Microbiol 2004; 42: 4242-49.

[56]   Maiden MCJ, Bygraves JA, Feil E, *et al.* Multilocus sequence typing: a portable approach to the identification of clones within populations of pathogenic microorganisms. Proc Natl Acad Sci USA 1998; 95: 3140-45.

[57]   Spratt, B. G. Multilocus sequence typing: molecular typing of bacterial pathogens in an era of rapid DNA sequencing and the Internet. Curr Opin Microbiol 1999; 2: 312-16.

[58]   Laciar A, Vaca L, Lopresti R, *et al.* DNA fingerprinting by ERIC-PCR for comparing Listeriasp. strains isolated from different sources in San Luis, Argentina. Rev Argent Microbiol 2006; 38: 55-60.

[59]   Kwon GH, Lee HA, Park JY, *et al.* Development of a RAPD-PCR method for identification of Bacillus species isolated from Cheonggukjang. Int J Food Microbiol 2009; 129: 282-87.

[60]   Filliol I, Sola C, Rastogi N. Detection of a previously unamplified spacer within the DR locus of Mycobacterium tuberculosis: epidemiological implications. J Clin Microbiol 2000; 38: 1231-34.

[61]   Niemann S, Richter E, Rüsch-Gerdes S. Differentiation among members of the Mycobacterium tuberculosis complex by molecular and biochemical features: evidence for two pyrazinamide-susceptible subtypes of M. bovis. J Clin Microbiol 2000; 38: 152-57.

[62]   Roring S, Hughes MS, Skuce RA, Neill SD. Simultaneous detection and strain differentiation of Mycobacterium bovis directly from bovine tissue specimens by spoligotyping. Vet Microbiol2000; 74: 227-36.

[63]   Bomfim MRQ, Bomfim MCK. Evaluation of LSSP-PCR for identification of Leptospirasp. In urine samples of cattle with clinical suspicion of leptospirosis. Vet Microbiol 2006; 118: 278-88.

[64]   Gordon SV, Brosch R, Billault A, *et al.* Identification of variable regions in the genomes of tubercle bacilli using bacterial artificial chromosome arrays. Mol Microbiol 1999; 32: 643-55.

[65]   Soini H, Pan X, Amin A, *et al.* Characterization of Mycobacterium tuberculosis isolates from patients in Houston, Texas, by spoligotyping. J Clin Microbiol 2000; 38: 669-76.

[66]   Bricker, BJ. PCR as a diagnostic tool for brucellosis. Vet Microbiol 2002; 90: 435-46.

[67]   Osterman B, Moriyon I. 2006. International Committee on Systematics of Prokaryotes. Subcommittee on the taxonomy of Brucella. Minutes of the meeting, 17 September 2003, Pamplona, Spain. Int J Syst Evol Microbiol 2006; 56: 1173-75.

[68]   Foster G, Osterman BS, Godfroid J, *et al.* Brucella ceti sp. nov. and Brucella pinnipedialis sp. nov. for Brucella strains with cetaceans and seals as their preferred hosts. Int J Syst Evol Microbiol 2007; 57: 2688-93.

[69]   Scholz HC, Hubalek Z, Sedlácek I, *et al.* Brucella microti sp. nov. , isolated from the common vole Microtus arvalis. Int J Syst Evol Microbiol 2008; 58: 375-82.

[70]   Scholz HC, Nöckler K, Göllner C, *et al.* Brucella inopinata sp. nov. , isolated from a breast implant infection. Int J Syst Evol Microbiol 2010; 60: 801-8.

[71]   Hinic V, Brodard I, Thomann A, *et al.* Novel identification and differentiation Brucella melitensis, B. abortus, B. suis, B. ovis, B. canis, and B. neotomae suitable for both conventional and real-time PCR systems. J Microbiol Meth 2008; 75: 375-78.

[72]   Huber, B, Scholz HC, Lucero N, Busse H. Development of a PCR assay for typing and subtyping of Brucella species. Int J Med Microbiol 2009; 299: 563-73.

[73]   Fekete A, Bantle JA, Halling SM, Sanborn MR. Preliminary development of a diagnostic test for Brucella using polymerase chain reaction. J Appl Bacteriol 1990; 69: 216-27.

[74]   Romero C, Gamazo C, Pardo M, Lopez-Goni I. Specific detection of Brucella DNA by PCR. J Clin Microbiol 1995; 33: 615-17.

[75]   Da Costa M, Guillou JP, Garin-Bastuji B, *et al.* Specificity of six gene sequences for the detection of the genus Brucella by DNA amplification. J Appl Bacteriol 1996; 81: 267-75.

[76]   García-Yoldi D, Marín CM, de Miguel MJ, *et al.* Multiplex PCR assay for the identification and differentiation of all Brucella species and the vaccine strains Brucella abortus S19 and RB51 and Brucella melitensis Rev1. Clin Chem 2006; 52: 779-81.

[77]   Bricker BJ, Halling SM. Differentiation of Brucella abortus bv. 1, 2, and 4, Brucellamelitensis, Brucella ovis, and Brucella suis bv. 1 by PCR. J Clin Microbiol 1994; 32: 2660-66.

[78]   Bricker BJ, Halling SM. Enhancement of the Brucella AMOS PCR assay for differentiation of Brucella abortus vaccine strains S19 and RB51. J Clin Microbiol 1995; 1640-42.

[79]   Ocampo-Sosa AA, Agüero-Balbín J, García-Lobo JM. Development of a new PCR assay to identify Brucella abortus biovars 5, 6 and 9 and the new subgroup 3b of biovar 3. Vet Microbiol 2005; 110: 41-51.

[80]   Fretin D, Whatmore AM, Al Dahouk S, *et al.* Brucella suis identification and biovar typing by real-time PCR. Vet Microbiol 2008; 131: 376-85.

[81]   Mercier E, Jumas-Bilak E, Allardet-Servent A, *et al.* Polymorphism in Brucella strains detected by studying distribution of two short repetitive DNA elements. J Clin Microbiol 1996; 34: 1299-02.

[82]   Tcherneva E, Rijpens N, Naydensky C, Herman L. Repetitive element sequence based polymerase chain reaction for typing of Brucella strains. Vet Microbiol 1996; 51: 169-78.

[83]   Fekete A, Bantle JA, Halling SM, Stich RW. Amplification fragment length polymorphism in Brucella strains by use of polymerase chain reaction with arbitrary primers. J Bacteriol 1992; 174: 7778-83.

[84]   Tcherneva E, Rijpens N, Jersek B, Herman LM. Differentiation of Brucella species by random amplified polymorphic DNA analysis. J Appl Microbiol 2000; 88: 69-80.

[85]   Guarino A, Serpe L, Fusco G, *et al.* Detection of Brucella species in buffalo whole blood by gene-specific PCR. Vet Rec 2000; 147: 634-36.

[86]   Sreevatsan S, Bookout JB, Ringpis F, *et al.* A multiplex approach to molecular detection of Brucella abortus and/or Mycobacterium bovis infection in cattle. J Clin Microbiol 2000; 38: 2602-10.

[87]   Amin AS, Hamdy ME, Ibrahim AK. Detection of Brucella melitensis in semen using the polymerase chain reaction assay. Vet Microbiol 2001; 83: 37-44.

[88]   Cortez A, Scarcelli E, Soares RM, *et al.* Detection of Brucella DNA from aborted bovine foetuses by polymerase chain reaction. Aust Vet J 2001; 79: 500-01.

[89]   Tantillo G, Di Pinto A, Vergara A, Buonavoglia C. Polymerase chain reaction for the direct detection of Brucella sp. in milk and cheese. J Food Prot 2001; 64: 164-67.

[90]   Leyla G, Kadri G, Umran O. Comparison of polymerase chain reaction and bacteriological culture for the diagnosis of sheep brucellosis using aborted fetus sample. Vet Microbiol 2003; 93: 53-61.

[91]   O'Leary S, Sheahan M, Sweeney T. Brucella abortus detection by PCR assay in blood, milk and lymph tissue of serologically positive cows. Res Vet Sci 2006; 81: 170-76.

[92]   Keid LB, Soares RM, Vasconcellos SA, *et al.* A polymerase chain reaction for the detection of Brucella canis in semen of naturally infected dogs. Theriogenology 2007; 67: 1203-10.

[93]   Keid LB, Soares RM, Vasconcellos SA, *et al.* A polymerase chain reaction for detection of Brucellacanis in vaginal swabs of naturally infected bitches. Theriogenology 2007; 68: 1260-70.

[94]   Guerra, M. A. Leptospirosis. JAVMA 2009; 234: 472-476.

[95]   Bharti AR, Nally JE, Ricaldi JN, *et al.* Leptospirosis: a zoonotic disease of global importance. Lancet Infect Dis 2003; 3: 757-71.

[96]   Levett PN. Leptospirosis. Clin Microbiol Rev 2001; 14: 296-26.

[97]   Croda J, Ramos JG, Matsunaga J, Queiroz A, *et al.* Leptospira immunoglobulin-like proteins as a serodiagnostic marker for acute leptospirosis. J Clin Microbiol. 2007; 45: 1528-34.

[98]   Heinemann MB, Garciab JF, Nunesb CM, *et al.* Detection and differentiation of Leptospirasp. serovars in bovine semen by polymerase chain reaction and restriction fragment length polymorphism. Vet Microbiol 2000; 73: 261-67.

[99]   Léon A, Pronost S, Tapprest J, *et al.* Identification of pathogenic Leptospira strains in tissues of a premature foal by use of polymerase chain reaction analysis. J Vet Diagn Invest 2006; 18: 218-21.

[100]   Lilenbaum W, Varges R, Brandão FZ, *et al.* Detection of Leptospira sp. in semen and vaginal fluids of goats and sheep by polymerase chain reaction. Theriogenol 2008; 69: 837-42.

[101]   Woo TH, Patal BK, Smythe LD, *et al.* Identification of pathogenic Leptospira genospecies by continuous monitoring of fluorogenic hybridization probes during rapid-cycle PCR. J Clin Microbiol 1997; 35: 3140-46.

[102]   Savio ML, Rossi C, Fusi P, *et al.* Detection and identification of Leptospira interrogans serovars by PCR coupled with restriction endonuclease analysis of amplified DNA. J Clin Microbiol 1994; 23: 935-41.

[103]   Woodward MJ, Redstone JS. Differentiation of Leptospira serovars by the polymerase chain reaction and restriction fragment length polymorphism. Vet Rec 1993; 132: 325-26.

[104]   Branger C, Blanchard B, Fillonneau C, *et al.* Polymerase chain reaction assay specific for pathogenic Leptospira based on the gene hap1 encoding the hemolysis-associated protein-1. FEMS Microbiol Lett 2005; 243: 437-45.

[105]   Liu D, Lawrence ML, Austin FW, *et al.* PCR detection of pathogenic Leptospira genomospecies targeting putative transcriptional regulator genes. Can J Microbiol 2006; 52: 272-77.

[106]   Jouglard SDD, Simionatto S, Seixas FK, *et al.* Nested polymerase chain reaction for detection of pathogenic leptospires. Can J Microbiol 2006; 52: 747-52.

[107]   Richtzenhain LJ, Cortez A, Heinemann MB, *et al.* A multiplex PCR for the detection of Brucella sp. and Leptospirasp. DNA from aborted bovine fetuses. Vet Microbiol 2002; 87: 139-47.

[108]   Mérien F, Amouriaux P, Perolat P, *et al.* Polymerase chain reaction for detection of Leptospira sp. in clinical samples. J Clin Microbiol 1992; 30: 2219-24.

[109]   Magajevski FS, Girio RJS, Mathias LA, *et al.* Detection of Leptospira sp. In the semen and urine of bulls serologically reactive to Leptospira interrogans serovar hardjo. Braz J Microbiol 2005; 36: 43-47.

[110]   Faber NA, Crawford M, LeFebvre RB, *et al.* Detection of Leptospira species in the aqueous humor of horses with naturally acquired recurrent uveitis. J Clin Microbiol 2000; 38: 2731-33.

[111]   Parra A, García N, García A, *et al.* Development of a molecular diagnostic test applied to experimental abattoir surveillance on bovine tuberculosis. Vet Microbiol2008; 127: 315-24.

[112]   De la Rua-Domenech R, Goodchild AT, Vordermeier HM, *et al.* Ante mortem diagnosis of tuberculosis in cattle: a review of the tuberculin tests, γ-interferon assay and other ancillary diagnostic techniques. Res Vet Sci2006; 81: 190-10.

[113]   Taylor GM, Worth DR, Palmer S, *et al.* Rapid detection of *Mycobacterium bovis* DNA in cattle lymph nodes with visible lesions using PCR. BMC Vet Res 2007; 3: 12.

[114]   Harrington NP, Surujballi OP, Waters WR, Prescott J. Development and evaluation of a real-time reverse transcription-PCR assay for quantification of gamma interferon mRNA to diagnose tuberculosis in multiple animal species. Clin Vaccine Immunol 2007; 14: 1563-71.

[115]   Serrano-Moreno BA, Romero TA, Arriaga C, *et al.* High frequency of Mycobacterium bovis DNA in colostra fro tuberculous cattle detected by nested PCR. Zoonoses and Public Health 2008; 55: 258-66.

[116]   Figueiredo EES, Silvestre FG, Campos WN, *et al.* Identification of *Mycobacterium bovis* isolates by a multiplex PCR. Braz J Microbiol 2009; 40: 231-233.

[117]   Wards BJ, Collins DM, Lisle GW. Detection of *Mycobacterium bovis* in tissues by polymerase chain reaction. Vet Microbiol1995; 43: 227-40.

[118]   Liébana E, Aranaz A, Mateos A, *et al.* Simple and rapid detection of *Mycobacterium tuberculosis* complex organisms in bovine tissue samples by PCR. J Clin Microbiol 1995; 33: 33-36.

[119]   Bouwman SA, Brown TA. Comparison between silica-based methods for the extraction of DNA from human bones from 18th-19th century London. Ancient Biomolecules 2002; 4: 173-78.

[120]   Heginbotham ML, Magee JT, Flanagan PG. Evaluation of the Idaho Technology Lightcycler PCR for the direct detection of *Mycobacterium tuberculosis* in respiratory specimens. Int J Tuberc Lung Dis2003; 7: 78-83.

[121] Norby B, Bartlett PC, Fitzgerald SD, *et al.* The sensitivity of gross necropsy, caudal fold and comparative cervical tests for the diagnosis of bovine tuberculosis. J Vet Diagn Invest2004; 16: 126-31.

[122] Zumárraga MJ, Meikle V, Bernardelli A, *et al.* Use of touch-down polymerase chain reaction to enhance the sensitivy of *Mycobacterium bovis* detection. J Vet Diagn Invest2005; 17: 232-38.

[123] Barouni AS, Augusto CJ, Lopes MTP, *et al.* A pncA polymorphism to differentiate between *Mycobacterium bovis* and Mycobacterium tuberculosis. Mol Cell Probes 2004; 18: 167-70.

[124] Chimara E, Ferrazoli L, Leão SC. Mycobacterium tuberculosis complex differentiation using gyrB-restriction fragment length polymorphism analysis. Mem Inst Oswaldo Cruz 2004; 99: 745-48.

[125] Sreevatsan S, Escalante P, Pan X, *et al.* Identification of a polymorphic nucleotide in oxyR specific for *Mycobacterium bovis.* J Clin Microbiol 1996; 34: 2007-10.

[126] Haas WH, Schilke K, Brand J, *et al.* Molecular analysis of katG gene mutations in strains of *Mycobacterium tuberculosis* complex from Africa. Antimicrob Agents Chemother 1997, 41: 1601-03.

[127] Espinosa de los Monteros LE, Galán JC, Gutiérrez M, *et al.* Allele-specific PCR method based on pncA and oxyR sequences for distinguishing *Mycobacterium bovis* from Mycobacterium tuberculosis: intraspecific M. bovis pncA sequence polymorphism. J Clin Microbiol 1998; 36: 239-42.

[128] Del Portillo P, Thomas MC, Martinez E *et al.* Multi-primer PCR system for differential identification of mycobacteria in clinical samples. J Clin Microbiol 1996; 34: 324-28.

[129] Herrera EA, Segovia M. Evaluation of mtp40 genomic fragment amplification for specific detection of Mycobacterium tuberculosis in clinical specimens. J Clin Microbiol 1996; 34: 1108-13.

[130] Weil A, Plikaytis BB, Butler WR, *et al.* The mtp40 gene is not present in all strains of Mycobacterium tuberculosis. J Clin Microbiol 1996; 34: 2309-11.

[131] Cobos-Marín L, Montes-Vargas J, Rivera-Gutierrez S, *et al.* A novel multiplex-PCR for the rapid identification of *Mycobacterium bovis* in clinical isolates of both veterinary and human origin. Epidemiol Infect2003; 130: 485-90.

[132] Fitzmaurice J, Sewell M, Manso-Silván L, *et al.* Real-time polymerase chain reaction assays for the detection of members of the Mycoplasma mycoides cluster. N Z Vet J 2008; 56: 40-47.

[133] Manson-Silván L, Perrier X, Thiaucourt F. Phylogeny of the Mycoplasma mycoides cluster based on analysis of five conserved protein-coding sequences and possible implications for the taxonomy of the group. Int J Syst Evol Microbiol 2007; 57: 2247-58.

[134] Bashiruddin JB, Frey J, K€onigsson MH, *et al.* Evaluation of PCR systems for the identification and differentiation of Mycoplasma agalactiae and Mycoplasma bovis: A collaborative trial. Vet J 2005; 169: 268-75.

[135] Sibila M, Pieters M, Molitor T, *et al.* Current perspectives on the diagnosis and epidemiology of Mycoplasma hyopneumoniae infection. Vet J 2009; 181: 221-231.

[136] Raviv Z, Kleven, SH. The development of diagnostic real-time TaqMan PCRs for the four pathogenic avian mycoplasmas. Avian Dis 2009; 53: 103-07.

[137] Ewing L, Cookson KC, Phillips RA, *et al.* Experimental infection and transmissibility of Mycoplasma synoviae with delayed serological response in chickens. Avian Dis 1998; 42: 230-238.

[138] Kleven SH, Rowland GN, Kumar MC. Poor serologic response to upper respiratory infection with Mycoplasma synoviae in turkeys. Avian Dis 2001; 48: 719-23.

[139] Thomas A, Dizier I, Linden A, *et al.* Conservation of the uvrC gene sequence in *Mycoplasma bovis* and its use in routine PCR diagnosis. Vet J 2004; 168: 100-02.

[140] Cottew GS, Breard A, DaMassa AJ, *et al.* Taxonomy of the Mycoplasma mycoides cluster. Isr J Med Sci 1987; 23: 632-35.

[141] Thiaucourt F, Lorenzon S, David A, Breard A. Phylogeny of the Mycoplasma mycoides cluster as shown by sequencing of a putative membrane protein gene. Vet Microbiol 2000; 72: 251-68.

[142] Woubit S, Manso-Silván L, Lorenzon S, *et al.* A PCR for the detection of mycoplasmas belonging to the Mycoplasma mycoides cluster: Application to the diagnosis of contagious agalactia. Mol Cell Probes 2007; 21: 391-99.

[143] Monnerat MP, Thiaucourt F, Poveda JB, *et al.* Genetic and serological analysis of lipoprotein LppA in Mycoplasma mycoides subsp. mycoides LC and Mycoplasma mycoides subsp. capri. Clin Diag Lab Immunol 1999: 224-30.

[144] Maigre L, Citti C, Marenda M, *et al.* Suppression-subtractive hybridization as a strategy to identify taxon-specific sequences within the Mycoplasma mycoides cluster: design and validation of an *M. capricolum* subsp. capricolum-specific PCR assay. J Clin Microbiol 2008; 46: 1307-16.

[145] Le Grand D, Saras E, Blond D, *et al.* Assessment of PCR for routine identification of species of the Mycoplasma mycoides cluster in ruminants. Vet Res 2004; 35: 635-49.

[146] Gorton TS, Barnett MM, Gull T, *et al.* Development of real-time diagnostic assays specific for Mycoplasma mycoides subspecies mycoides Small Colony. Vet Microbiol 2005; 111: 51-58.

[147] Hotzel K, Sachse K, Pfeutzner H. Rapid detection of Mycoplasma bovis in milk samples and nasal swabs using the polymerase chain reaction. J Appl Bacteriol 1996; 80: 505-10.

[148] Bölske G, Mattsson JG, Bascuñana CR, *et al.* Diagnosis of contagious caprine pleuropneumonia by detection and identification of Mycoplasma capricolum subsp. capripneumoniae by PCR and restriction enzyme analysis. J Clin Microbiol 1996; 34: 785-91.

[149] Monnerat MP, Thiaucourt F, Nicolet J, Frey J. Comparative analysis of the lppA locus in Mycoplasma capricolum subsp. capricolum and Mycoplasma capricolum subsp. capripneumoniae. Vet Microbiol 1999; 69: 157-72.

[150] Vilei EM, Nicolet J, Frey J. IS1634, a novel insertion element creating long, variable-length direct repeats which is specific for Mycoplasma mycoides subsp. mycoides small-colony type. J Bacteriol 1999; 181: 1319-23.

[151] Miles K, Churchward CP, McAuliffe L, *et al.* Identification and differentiation of European and African/Australian strains of Mycoplasma mycoides subspecies mycoides small-colony type using polymerase chain reaction analysis. J Vet Diagn Invest 2006; 18: 168-71.

[152] Director General de Producción Agropecuaria. Resolución de 24 de mayo de 2007, por la que desarrollan los programas sanitarios a realizar por las Agrupaciones de Defensa Sanitaria Ganaderas, definidos en la Orden AYG/1131/2006, de 30 de junio. Comunidad Autónoma de Castilla y Leon, Spain, 2006.

[153] De la Fe C, Martín AG, Amores J, *et al.* Latent infection of male goats with Mycoplasma agalactiae and Mycoplasma mycoides subspecies capri at an artificial insemination centre. Vet J 2010; 186: 113-115.

[154] Chávez González YR, Bascuñana CR, Bölske G, *et al. In vitro* amplification of the 16S rRNA genes from Mycoplasma bovis and Mycoplasma agalactiae by PCR. Vet Microbiol 1995; 47: 183-90.

[155] Johansson KE, Heldtander MU, Pettersson B. Characterization of mycoplasmas by PCR and sequence analysis with universal 16S rDNA primers. Methods Mol Biol 1998; 104: 145-65.

[156] Königsson MH, Bölske G, Johansson KE. Intraspecific variation in the 16S rRNA gene sequences of Mycoplasma agalactiae and Mycoplasma bovis strains. Vet Microbiol 2002; 85: 209-20.

[157] Subramaniam S, Bergonier D, Poumarat F, *et al.* Species identification of Mycoplasma bovis and Mycoplasma agalactiae based on the uvrC genes by PCR. Mol Cell Probes 1998; 12: 161-69.

[158] Foddai A, Idini G, Fusco M, *et al.* Rapid differential diagnosis of Mycoplasma agalactiae and Mycoplasma bovis based on a multiplex-PCR and a PCR-RFLP. Mol Cell Probes 2005; 19: 207-12.

[159] Lorusso A, Decaro N, Greco G, *et al.* A real-time PCR assay for detection and quantification of Mycoplasma agalactiae DNA. J Appl Microbiol 2007; 103: 918-23.

[160] Fitzmaurice J, Sewell M, King CM, *et al.* A real-time polymerase chain reaction assay for the detection of Mycoplasma agalactiae. N Z Vet J 2008; 56: 233-36.

[161] Cai HY, Bell-Rogers P, Parker L, Prescott JF. Development of a real-time PCR for detection of Mycoplasma bovis in bovine milk and lung samples. J Vet Diagn Invest. 2005; 17: 537-45.

[162] Cremonesi P, Vimercati C, Pisoni G, *et al.* Development of DNA extraction and PCR amplification protocols for detection of Mycoplasma bovis directly from milk samples. Vet Res Commun 2007; 31: 225-27.

[163] McDonald WL, Rawdon TG, Fitzmaurice J, *et al.* Survey of bulk tank milk in New Zealand for Mycoplasma bovis , using species-specific nested PCR and culture. N Z Vet J 2009; 57: 44-49.

[164] Pinnow CC, Butler JA, Sachse K, *et al.* Detection of Mycoplasma bovis in preservative-treated fi eld milk samples. J Dairy Sci 2001; 84: 1640-45.

[165] Caron J, Ouardani M, Dea S. Diagnosis and differentiation of Mycoplasma hyopneumoniae and Mycoplasma hyorhinis infections in pigs by PCR amplification of the p36 and p46 genes. J Clin Microbiol 2000; 38: 1390-96.

[166] Kurth KT, Hsu T, Snook ER, *et al.* Use of a Mycoplasma hyopneumoniae nested polymerase chain reaction test to determine the optimal sampling sites in swine. J Vet Diagn Invest 2002; 14: 463-69.

[167] Cai HY, van Dreumel T, McEwen B, *et al.* Application and field validation of a PCR assay for the detection of Mycoplasma hyopneumoniae from swine lung tissue samples. J Vet Diagn Invest 2007; 19: 91-95.

[168] Hammond PP, Ramírez AS, Morrow CJ, Bradbury JM. Development and evaluation of an improved diagnostic PCR for Mycoplasma synoviae using primers located in the haemagglutinin encoding gene vlhA and its value for strain typing. Vet Microbiol 2009; 136: 61-68.

[169] Garcia M, Jackwood MW, Head M, *et al.* Use of species-specific oligonucleotide probes to detect Mycoplasma gallisepticum, M. synoviae and M. iowae PCR amplification products. J Vet Diagn Investig 1996; 8: 56-63.

[170] Jarquin R, Schultz J, Hanning I, Ricke SC. Development of a real-time polymerase chain reaction assay for the simultaneous detection of Mycoplasma gallisepticum and Mycoplasma synoviae under industry conditions. Avian Dis 2009; 53: 73-77.

[171] Ramírez AS, Naylor CJ, Hammond PP, Bradbury JM. Development and evaluation of a diagnostic PCR for Mycoplasma synoviae using primers located in the intergenic spacer region and the 23S rRNA gene. Vet Microbiol 2006; 118: 76-82.

[172] Bencina D, Drobnic-Valic M, Horvat S, *et al.* Molecular basis of the length variation in the N-terminal part of Mycoplasma synoviae hemagglutinin. FEMS Microbiol Lett 2001; 203: 115-23.

[173] Hong Y, García M, Leiting V, *et al.* Specific detection and typing of Mycoplasma synoviae strains in poultry with PCR and DNA sequence analysis targeting the hemagglutinin encoding gene vlhA. Avian Dis 2004; 48: 606-16.

[174] Mardassi BB, Mohamed RB, Gueriri I, *et al.* Duplex PCR to differentiate between Mycoplasma synoviae and Mycoplasma gallisepticum on the basis of conserved species-specific sequences of their hemagglutinin genes. J Clin Microbiol 2005; 43: 948-58.

[175] Jeffery N, Gasser RB, Steer PA, Noormohammadi AH. Classification of Mycoplasma synoviae strains using single-strand conformation polymorphism and high-resolution melting-curve analysis of the vlhA gene single copy region. Microbiol 2007; 153: 2679-88.

[176] Salyers AA, Whitt DD. Staphylococcus species. In: Salyers AA, Whitt DD, Eds. Bacterial pathogenesis - A molecular approach. Washington, ASM Press, 2002; pp. 216-31.

[177] Studer E, Schaerent W, Naskova J, *et al.* A longitudinal field study to evaluate the diagnostic properties of a quantitative real-time polymerase chain reaction-based assay to detect Staphylococcus aureus in milk. J Dairy Sci 2008; 91: 1893-02.

[178] Taponen S, Salmikivi L, Simojoki H, *et al.* Real-time polymerase chain reaction-based identification of bacteria in milk samples from bovine clinical mastitis with no growth in conventional culturing. J Dairy Sci 2009; 92: 2610-17.

[179] Sears PM, Smith BS, English PB, *et al.* Shedding pattern of Staphylococcus aureus from bovine intramammary infections. J Dairy Science 1990; 73: 2785-89.

[180] Riffon R, Sayasith K, Khalil H, *et al.* Development of a rapid and sensitive test for identification of major pathogens in bovine mastitis by PCR. J Clin Microbiol 2001; 39: 2584-89.

[181] Phuektes P, Mansell PD, Browning GF. Multiplex polymerase chain reaction assay for simultaneous detection of Staphylococcus aureus and streptococcal causes of bovine mastitis. J Dairy Sci 2001; 84: 1140-48.

[182] Bialkowska-Hobrzanska H, Harry HV, Jaskot D, Hammerberg O. Typing of coagulase-negative staphylococci by southern hybridization of chromosomal DNA fingerprints using a ribosomal RNA probe. Eur J Microbiol Infect Dis 1990; 9: 588-94.

[183] Vannuelfel P, Heusterspreute M, Bouyer M, *et al.* Molecular characterization of femA from Staphylococcus hominis and Staphylococcus saprophyticus and femA-based discrimination of staphylococcal species. Res Microbiol 1999; 150: 129-41.

[184] Poyart C, Quesne G, Boumaila C, Trieu-Cuot P. Rapid and accurate species-level identification of coagulase-negative staphylococci by using the sodA gene as a target. J Clin Microbiol 2001; 39: 4296-01.

[185] Drancourt M, Raoult D. rpoB gene sequence-based identification of Staphylococcus species. J Clin Microbiol 2002; 40: 1333-38.

[186] Hein I, Jørgensen HJ, Loncarevic S, Wagner M. Quantification of Staphylococcus aureus in unpasteurised bovine and caprine milk by real-time PCR. Res Microbiol 2005; 156: 554-63.

[187] Bannoehr J, Franco A, Iurescia M, *et al.* Molecular diagnostic identification of *staphylococcus pseudintermedius*. J Clin Microbiol 2009; 47: 469-71.

[188] Cai HY, Archambault M, Gyles L, Prescott JF. Molecular genetic methods in the veterinary clinical bacteriology laboratory: current usage and future applications. Anim Health Res Rev 2003; 4: 73-93.

CHAPTER 5

PCR Detection of Viruses in Veterinary Medicine

Yihang Li[1,*], Sudhir K. Ahluwalia[2] and Mark D. Freeman[3]

[1]*Geriatrics Research, Education and Clinical Center (GRECC), Department of Veterans, Affairs Palo Alto Health Care System, Palo Alto, California 94304;* [2]*Banfield Pet Hospital, 10 Traders Way, Salem, MA 01970 and* [3]*Ross University School of Veterinary Medicine, P. O. Box 334, Basseterre, St. Kitts, West Indies 00265*

Abstract: Since the invention of PCR by Michael Smith and Kary Mullis, the last two decades have seen an explosion of PCR application in various aspects of biological and medical sciences. Due to its high sensitivity, versatility and reproducibility, PCR has become one of the standard procedures in diagnosis of almost all viral diseases in veterinary medicine. Unlike serological methods, which rely on the presence of specific antibodies, and may lead to false positive or false negative results, PCR detects the presence or absence of the pathogen, thereby providing a better measure of the viruses. Furthermore, with the advent of real-time PCR, researchers now can quantify the amount of virus that is present at different sites in the animal, thereby gaining the ability to determine the stage of infection. Variations of PCRs also allow phenotypic characterization between different viral isolates and between wild-type viruses and vaccines, while allowing simultaneous diagnosis of multiple viruses. PCR has become one of the most commonly used methods in diagnosis of viral disease in livestock and companion animals, and with the development of automated technologies and multiplex PCR systems, which vastly elevate the throughput of PCR assays, increased use of PCR-based techniques is expected in the future.

Keywords: Polymerase chain reaction; molecular diagnosis; adenovirus; calicivirus; coronavirus; flavivirus; herpesvirus; orthomyxovirus; paramyxovirus; parvovirus; retrovirus.

1. INTRODUCTION

Rapid and accurate determination of the etiology of viral infections is critical for effective treatment and prophylaxis in veterinary practices. However, traditional diagnostic assays such as cell culture, serology, and antigen detection assays are typically costly, time-consuming, and tedious. Diagnosis of infectious diseases in animals, including viral diseases, has been revolutionized as molecular techniques, particularly PCR and RT-PCR, become increasingly accessible [1]. In recent years, the application of PCR techniques has matured sufficiently for not only the detection of viruses, but also epidemiological assessment, phenotypic evaluation, drug resistance tests, *etc* [2, 3]. Here, the authors attempt to review the universal issues of applying the PCR technology in animal viral diseases, followed by representative examples of how PCR and RT-PCR is used for most common viral agents in veterinary medicine.

1.1. Sample Processing

Almost all sample types can be used for PCR assays. Appropriate sample type depends on the virus, and the most commonly used sample types are blood, serum, urine, and other bodily fluids. For example, in routine isolation of Newcastle disease virus from chickens, turkeys, and other birds, samples are obtained by swabbing the trachea and the cloaca, whereas for feline infectious peritonitis virus, pleural effusion would be the optimum sample source. Different extraction techniques of variable complexity are available. The product of simply boiling the sample to release the RNA from the virus has been assayed [4]. Direct coating of the viral suspension to plastic wells has also been assayed, as it is supposed that the structural distortion of the virion, due to plastic adherence, favors accessibility to the RNA [5]. Most laboratories use the rapid acid-guanidinium thiocyanate method, described by Chomczynski and Soachi, which is now commercially available from most major companies [6].

***Address correspondence to Yihang Li:** Geriatrics Research, Education and Clinical Center (GRECC), Department of Veterans Affairs Palo Alto Health Care System, Room C315, Building 4, 3801 Miranda Avenue, Palo Alto, California 94304, USA; Tel: 650-493-5000x66344; Fax:650-496-2505; E-mail: liyihang@stanford.edu*

1.2. PCR Types

The inception of quantitative PCR (or real-time PCR) has gradually replaced the application of regular PCRs. Generally speaking, quantitative PCRs have the following advantages over traditional PCRs: i) it gives the copy number of target genes (which are often diagnostically critical), not just "positive" and "negative"; ii) it generally uses shorter amplicons, therefore PCR is faster; iii) it allows higher throughput due to ease of operation (no need for post-PCR assays, such as agarose gels); iv) it allows multiplex assays to detect several pathogens simultaneously [7]. Four major types of quantitative PCRs have been developed: TaqMan®, Molecular Beacons, Scorpions® and SYBR Green®, each of which have very broad applications. More information can be found at Chapter **3** of this book.

Nested PCR. Nested PCR is a modification of PCR intended to reduce the misamplification in products due to the amplification of unexpected primer binding sites. It typically involves two sets of primers used in two rounds of PCR, with the second set intended to amplify a region within the first amplicon. Nested PCR is used in diagnosis of many viral pathogens.

Multiplex PCR. Multiplex PCR is a variant of PCR that simultaneously amplifies multiple DNA targets using more than one pair of primers in a single reaction. It has been successfully used in detection, typing and subtyping of viruses. Multiplex is widely used for simultaneous detection and typing of viruses. For example, Baxi *et al.* used common primers and type-specific (BVDV1 and BVDV2) TaqMan probes to detect as low as 10-100 $TCID_{50}$ (Tissue culture infectious dose 50%) of virus, at the same time typing BVDV strains and field isolates [8]. Thus, the one-step real-time RT-PCR assay appears to be a rapid, sensitive, and specific test for detection and typing of BVDV. Caterina *et al.* developed a multiplex PCR assay that allows simultaneous detection and differentiation of avian reovirus (ARV), avian adenovirus group I (AAV-I), infectious bursal disease virus (IBDV), and chicken anemia virus (CAV), and greatly improved the spectrum of application of PCR in viral agent detection [9]. It is also possible to have multiplex PCR assays that can detect viral and bacterial pathogens simultaneously. For example, Sykes *et al.* developed a multiplex RT-PCR that can identify feline herpesvirus 1 (FHV1), feline calicivirus (FCV) and *Chlamydia psittaci* in cats with upper respiratory tract disease [10].

Immuno-PCR. Reuter *et al.* used an immune-real-time PCR to quantitatively determine the presence of prions [11]. They used a direct conjugate of a prion-specific antibody and a synthetic DNA tail, which allows subsequent quantification of restricted DNA tails using real-time PCR. The assay was used to detect scrapie prions bound to polyvinylidene difluoride membranes and could detect scrapie prions with high sensitivity, showing great potential for indirect PCR detection of transmissible spongiform encephalopathies (TSEs).

1.3. Other Applications of PCR in Animal Diseases

PCR has been used for applications other than diagnosis of viral diseases. For example, Ohe *et al.* have used PCR for phylogenetic analysis of many veterinary viral agents, such as caliciviruses and poxviruses [12, 13]. Hans-Peter Ottiger developed standardized PCR assays for multiple avian viral agents, including avian leucosis virus, avian orthoreovirus, infectious bursal disease virus, infectious bronchitis virus, Newcastle disease virus, infectious laryngotracheitis virus, influenza A virus, Marek's disease virus, turkey rhinotracheitis virus, egg drop syndrome virus, chicken anemia virus, avian adenovirus and avian encephalomyelitis virus, to test the purity of avian viral vaccines [14]. As an alternative to traditional serological methods or viral culture, PCR provides a significantly higher sensitivity and degree of discrimination between different viruses.

1.4. Causes of False Results

Use of positive and negative controls is critical for correct interpretation of PCR results. Ideally, a range of positive controls, from strongly positive (as abundant as in acutely infected animals), to weakly positive (just above the detection limit of the PCR assay) should be used [2]. As PCR assays increase the sensitivity of viral detection by degrees of magnitude, the most common cause of false positive results is contamination, especially in detection of RNA viruses. Frequently disinfecting the workbench, proper handling of PCR reagents, clean nucleic acid extraction, PCR amplification and post-PCR manipulation is therefore recommended. Some researchers have introduced extra steps in the process, to ensure that false positive signals are not picked up. For example, Nuanualsuwan *et al.* pretreated the samples with proteinase K and ribonuclease to effectively prevent positive results by inactivated vaccines [15]. Alternatively, because inactivated FIV vaccine virus does not integrate its RNA into the host genome, a PCR assayed target at the *gag* gene was developed and discriminates between infected and vaccinated cats [16]. Also, it is important to keep the same routine in processing repeated samples to keep data consistent, so that quantitative results are comparable across different batches of processed samples.

2. PCR DETECTION OF VIRUSES IN VETERINARY MEDICINE

Currently PCR is used in diagnosis of most viral agents. In the following, some of the most important viruses in veterinary practice are listed. A more comprehensive list of PCR detection of viruses in animals can be found in Table **1** at the end of this chapter.

2.1. Adenoviruses

Adenoviruses are Class I (double stranded DNA) viruses, non-enveloped, icosahedral viruses that are typically 74-80 nm in diameter. Their linear genomes are approximately 25-45 kb and code for about 20 proteins [17]. There are five genera in the family Adenoviridae: Aviadenoviruses, Atadenovirus, Ichtadenovirus, Mastadenovirus, and Siadenovirus (http://www.ncbi.nlm.nih.gov/taxonomy). Aviadenoviruses are pathogens of fowl , whereas two types of canine adenoviruses in the Genus Mastadenovirus, cause infectious canine hepatitis and infectious tracheobronchitis, respectively, in dogs [18].

Avian Adenovirus. Avian adenoviruses, or aviadenoviruses, are a diverse group of adenoviruses that affect parrots, pigeons, turkeys and chickens. Depending on the infected species, their disease manifestations include enteritis, splenitis, inclusion body hepatitis, bronchitis, pulmonary congestion ventriculitis, pancreatitis, and edema [19, 20]. Diagnosis of avian adenovirus was usually performed by isolation of virus followed by serological typing methods, histopathology, electron microscopy, viral isolation and ELISA [21]. Raue *et al.* first employed PCR to detect the avian adenovirus, using the hexon gene as the target, in chickens and turkeys [22, 23]. Combining PCR with subsequent restriction enzyme analysis and/or product sequencing allows identification of multiple strains and serotypes [22, 24]. These works were followed by Ganesh and colleagues, who then used PCR to identify avian adenoviruses associated with hydropericardium hepatitis syndrome [25]. With the advent of real-time PCR, quantitative PCR was also used in the detection of avian adenoviruses in different organs of the fowls, thus greatly improving the diagnostic and research values of PCR [26-28].

Canine Adenovirus Type 1. Canine adenovirus Type I (CAV-1) causes infectious canine hepatitis, an acute liver infection in dogs [29]. It also causes disease in wolves, coyotes, and bears, and encephalitis in foxes. The virus is contracted through the mouth or nose, and is spread in the feces, urine, blood, saliva, and nasal discharge of infected dogs. The virus can cause fever, depression, anorexia, coughing, and possibly corneal edema and signs of liver disease, such as jaundice, vomiting, and hepatic encephalopathy. Severe cases will develop bleeding disorders, which can cause hematomas to form in the mouth, or even death [30, 31]. Diagnosis of CAV-1 is typically dependent on recognizing the combination of symptoms and abnormal blood tests that occur in infectious canine hepatitis, or a rising antibody titer to CAV-1 that can be detected by ELISA [32]. PCR made it possible to detect CAV-1 with high sensitivity and specificity, differentiating it from canine parvovirus, which can cause highly similar symptoms (see 2.7), and it is very difficult to distinguish between CAV-1 and canine adenovirus Type 2 (CAV-2) by haemagglutination and neutralization tests. Kiss *et al.* first utilized PCR for diagnosis of CAV-1 [33], followed by Hu *et al.,* who used different amplicon lengths to differentiate CAV-1 and CAV-2 infections [34, 35]. Differentiation of CAV-1 and canine parvovirus is also reliably achieved [36].

2.2. Caliciviruses

Caliciviruses are Class IV (positive-sense, single stranded RNA) viruses that are found in a wide range of mammals. This family has four recognized genera: *Vesivirus, Lagovirus, Norovirus* and *Sapovirus*. Notable disease agents of this family include Feline calicivirus (FCV), murine norovirus 1 (MNV-1) and Rabbit hemorrhagic disease virus (RHDV).

Feline Calicivirus. The most important species of the family Caliciviridae is FCV, a virus of the genus *Vesivirus* that infects cats. It is one of the two important viral causes of respiratory infection in cats, the other being feline herpesvirus (see 2.4). FCV can be isolated from about 50 percent of cats with upper respiratory infection [37]. Different strains of FCV have variable levels of virulence, causing a range of clinical syndromes from inapparent infections to relatively mild oral and upper respiratory tract disease with or without acute lameness [38]. As an RNA virus, the genome of FCV is highly variable, making it highly adaptable to environmental pressures. This not only allows for the development of more virulent strains, but also creates problems for detection of the virus [39-41]. Before PCR, diagnosis of FCV was usually done by virus culture and immunohistochemical staining, but it could be difficult due to the fact that the FCV symptoms are often similar to other feline respiratory diseases, especially feline viral rhinotracheitis, and there are frequent co-infections [39]. Application of PCR greatly increased the specificity of FCV diagnosis.

One form of FCV has been found to cause a particularly severe systemic disease in cats, similar to rabbit hemorrhagic disease (caused by rabbit hemorrhagic disease virus, also a calicivirus). This virus has been called virulent systemic feline calicivirus (VS-FCV) or FCV-associated virulent systemic disease (VSD). Several groups developed a RT-PCR assay to amplify a hypervariable region of the FCV capsid gene. The sequence from this region was used to compare viruses used in three attenuated vaccines to viruses isolated from vaccinated cats with clinical signs of FCV-infection [42-44]. Radford *et al.* also compared RT-PCR with serology-based detection methods, and proved that the sequence-based method is preferable [45].

2.3. Coronaviruses

Coronaviruses are also Class IV (positive-sense, single stranded RNA) viruses. Notable members of the family Coronaviridae include SARS (severe acute respiratory syndrome) virus which caused a worldwide pandemic in 2003 (http://www.who.int/csr/sars/en), avian infectious bronchitis virus, bovine coronavirus (BCV), canine coronavirus (CCV), feline coronavirus virus (FCV, a mutation of which results in feline infectious peritonitis or FIP virus), turkey coronavirus (TCV, or bluecomb virus), and porcine epidemic diarrhea virus (PEDV). Many economically important diseases in veterinary science and animal husbandry are attributed to coronaviruses. Avian infectious bronchitis virus (IBV), the first coronavirus to be isolated, causes an acute and highly infectious respiratory disease in chickens [46, 47]. Coronavirus primarily causes respiratory symptoms but a few strains can also lead to other diseases, among which enteritis is by the far the most common disease produced by coronavirus in a variety of hosts. For example, porcine transmissible gastroenteritis virus (TGEV) causes severe disease in piglets with mortality up to 100% [48, 49]. Porcine respiratory coronavirus (PRCV) is a close relative of TGEV but causes only mild respiratory disease [50].

Canine Coronavirus (CCV). CCV is a member of the genus Alphacoronavirus. There are two canine coronaviruses, a Group I coronavirus called canine enteric coronavirus (CECoV), and a Group II coronavirus called canine respiratory coronavirus (CRCoV) [51, 52]. The two viruses are genetically and antigenically distinct. CRCoV was initially discovered in dogs with acute respiratory infection in England in 2003 [53, 54], and is genetically related to the bovine and human coronaviruses, whereas CECoV is more related to the feline coronavirus [53, 55]. CECoV is mainly responsible for diarrhea in young animals and in a few cases can also cause vomiting. CRCoV is part of a complex of bacteria and viruses associated with kennel cough or CIRD (Canine infectious respiratory disease). Areas with increased animal density such as kennels, veterinary hospitals, shelters *etc.*, increase the risk of infection. Definitive diagnosis of CCV can be made by electron microscopy, and immunohistochemistry can be used to demonstrate the viral antigen as well. The PCR assays offered by diagnostic laboratories are most sensitive and specific for diagnosing coronavirus infection [56, 57].

Feline Coronavirus (FCV). FCV causes feline infectious peritonitis (FIP), a common and deadly disease in cats that is difficult to diagnose ante-mortem due to lack of any pathognomonic change or specific clinical signs [58, 59]. In veterinary practice, a weighted score system based on various clinical signs and laboratory changes is commonly used, but this scoring system is not of diagnostic importance [60]. An increase in total serum protein is the most consistent laboratory finding in cats suffering from FIP, but serum albumin to globulin ratio has a higher diagnostic importance. Meanwhile, cerebrospinal fluid analysis may reveal 2-7 fold increases in protein along with pleocytosis. However, many cats with neurological signs also have normal CSF values, therefore routine tests are often low in sensitivity and specificity [61].

FIP virus is a genetic mutant of feline enteric coronavirus, and the mutations can occur in numerous potential sites and are not well understood. This limits the ability of PCR to distinguish between mutated pathogenic FIP virus and non-mutated enteric corona virus [62]. In addition, since feline coronavirus is found in the blood stream of healthy cats, positive PCR results from blood should be critically evaluated. PCR results are highly sensitive to detect Feline coronavirus in fecal samples and can be used to detect chronic shedders, however positive PCR in a fecal sample does not indicate that a cat has, or is more predisposed to, developing FIP [63].

As per current understanding of pathology of FIP, the enteric coronavirus should not only not cross the intestinal barrier to reach the blood stream but should also not acquire the ability to replicate in the mononuclear cells such as macrophages. Therefore, detecting and quantifying the replicating virus outside the intestinal tract can diagnose FIP with high specificity [64]. The peculiar replicating cycle of coronavirus produces subgenomic RNA of all of its genes, which is produced only during viral replication. Its presence in the blood stream, ascitic fluid, CSF or any other tissue (except the intestinal tract) can be correlated with FIP with statistical significance. The Auburn University Molecular Diagnostics Laboratory (www.vetmed.auburn.edu/molecular-diagnostics) has used this approach to target the subgenomic mRNA of the M gene of coronavirus, and this RT-PCR specifically detects and quantifies the replicating virus as opposed to only detecting the presence of viral genomic RNA that may or may not be associated with active viral replication (unpublished data).

2.4. Flavivirus

Flaviviruses are also Class IV (positive-sense, single stranded RNA) viruses. With three genera (hepacivirus, flavivirus and pestivirus), they include Dengue Fever virus, Yellow Fever virus, Hepatitis C virus, Classical Swine Fever (CSF), Border Disease Virus (BDV) and Bovine Viral Diarrhea Virus (BVDV) [65].

Bovine viral diarrhea is a widespread disease of cattle caused by BVDV. The disease often causes reduced growth , milk production, and reproduction performance, and increased susceptibility to other diseases in different developmental stages of cattle [66]. There are two genotypes of BVDV, BVDV1 and BVDV2, which are antigenically related. Frequently nonhomologous RNA recombination events can lead to the emergence of genetically distinct strains that are lethal to the host. Because of the immunosuppressive effect of BVDV, the persistently infected (PI) animals are often co-infected by other viruses or bacteria, thereby relating BVDV to numerous respiratory, enteric, immune and reproductive diseases [67]. Because of the huge economic losses, control (most importantly by vaccination) of BVDV is critical to the cattle industry [68-70].

Preliminary diagnosis of BVDV is done by a variety of methods, most routinely by virus isolation. Since the level of viremia in PI animals is generally very high (up to 10^7 CCID$_{50}$, or cell culture infectious dose 50% per ml serum), ear notch biopsies are obtained for visualization of the virus under the microscope. An immunohistochemistry (IHC) test using fixed skin biopsies is widely used in diagnosis of PI animals [71]. An antigen capture ELISA test (ACE) is also available. Gel-based and real time PCR tests have also been applied to detect BVDV genomic material in samples of fluids from PI cattle. In 1991, several groups started to use PCR to detect BVDV [72, 73], followed by other clinicians and researchers [74]. With a multiplex real-time PCR assay Baxi *et al.* could detect and type the BVDV strains with high sensitivity and reproducibility [8].

2.5. Herpesviruses

Herpesviruses are Class I (double stranded DNA) viruses that infect humans and other mammals, including cows, horses, dogs, cats, monkeys, and birds. Based on genome organization, herpesviruses are divided into three subfamilies, Alpha-, Beta- and Gammaherpesvirinae. Most clinically important herpesviruses belong to the subfamily Alphaherpesvirinae.

Avian Herpesvirus. Avian herpesvirus causes Marek's Disease, a highly contagious viral neoplastic disease in chickens. Therefore, it is also called Marek's Disease Virus (MDV) or gallid herpesvirus 2 (GaHV-2). Marek's Disease, named after the Hungarian veterinarian Dr. József Marek, can have acute or chronic symptoms, including asymmetric paralysis of one or more limbs, lesions in the peripheral nerves, lymphomatous infiltration/tumors in the skin, skeletal muscle, and visceral organs, atherosclerosis, and immunosuppression [75-77]. Its ocular form can cause lymphocyte infiltration of the iris, anisocoria, and blindness, whereas the cutaneous form causes round, firm lesions at the feather follicles [37].

Vaccines made of attenuated viruses or low virulence strains have been used since 1968, and have prevented large economic losses worldwide. However, the practice is controversial. Some researchers argue that the use of vaccines provide a driving force for the virus to evolve toward higher virulence, causing the efficacy of vaccines to decrease constantly [78]. At this stage, either novel vaccines need to be developed, or current vaccines need to be improved by optimization of timing and vaccine delivery route, as well as vaccination regimens specific for different breeds of chicken [76]. Enlargement of the ischiatic (sciatic) nerve along with suggestive clinical signs and the presence of nodules on the internal organs in a bird of 3-4 months of age is highly suggestive of Marek's Disease. Confirmation is usually made by histological demonstration of a lymphomatous infiltration, although it does need to be differentiated from lymphoid leukosis [76]. Sliva first developed a PCR assay that is capable of differentiating pathogenic and non-pathogenic variants of MDV serotype 1 [79], followed by Becker *et al.*, whose assay further distinguished between pathogenic and vaccine strains. Meanwhile, Wang *et al.* used the thymidine kinase gene as the amplification target, and obtained positive results in the bursae, kidneys and feathers of diseased chickens, and negative results with vaccinated chickens [80]. Reinmann used PCR to identity reticuloendotheliosis virus (REV), a frequent contaminant in MDV vaccines [81].

Bovine Herpesvirus. There are four major types of bovine herpesviruses, bovine herpesvirus 1 (BHV-1), bovine herpesvirus 2 (BHV-2), and bovine herpesvirus 4 (BHV-4) in the Alphaherpesvirinae subfamily, and bovine herpesvirus 5 (BHV-5) and alcelaphine herpesvirus 1 in the Gammaherpesvinae subfamily. BHV-1 can cause rhinotracheitis, vaginitis, balanoposthitis, fertility disorders, conjunctivitis, enteritis, shipping fever, and bovine respiratory disease complex (BRDC) [37, 82]. With bovine respiratory syncytial virus (BRSV), parainfluenzavirus-3 (PI3), bovine coronavirus, bovine viral diarrhea virus (BVDV) and bovine reovirus, bovine herpesvirus is one of the viruses that cause severe respiratory diseases, and hence large economic losses in the bovine industry [83]. Vaccines that are created with inactivated and modified live viruses that effectively reduce the severity and incidence of disease are available [84]. BHV-2 causes two diseases in cattle, bovine mammillitis and pseudo-lumpyskin disease. BHV-2 is similar in structure to human herpes simplex virus. The strain of BHV-2 that causes pseudo-lumpyskin disease is also known as the Allerton virus. Bovine herpesvirus is diagnosed by visualization of virus or virus components, serological tests, or by molecular methods such as PCR, nucleic acid hybridization and sequencing [84]. Vilcek first used PCR for detection of BHV-1, using the gl gene as the amplification target [85], followed by Liang *et al.* and Wiedmann [86, 87]. However, diagnosis of BHV-2 is often complicated by co-infections with bacterial or other viral agents. There was no report of diagnosing BHV-2 by PCR until 2002, when two groups developed multiplex PCR assays for the simultaneous detection of BHV-1 and BHV-2 [83, 88].

2.6. Orthomyxoviruses

Orthomyxoviruses are segmented Class V (negative sense single-stranded RNA) viruses. Of the five genera in the family of orthomyxoviridae (Influenzavirus A, Influenzavirus B, Influenzavirus C, Isavirus and Thogotovirus), Influenza virus A has received the most attention, especially after recent outbreaks of H5N1 avian influenza and H1N1 swine origin influenza pandemics in humans and other mammals (www.who.int/csr/disease/swineflu/en/index.html).

There is only one species in the genus Influenzavirus A, influenza virus A. However, multiple subtypes are identified based on two viral surface glycoproteins hemagglutinin (H or HA) and neuraminidase (N or NA). Sixteen H subtypes (or serotypes) and nine N subtypes have been identified, causing vast differences in host range and infectivity [89, 90]. Whereas all subtypes are found in birds (natural hosts), there are a limited number of mammalian subtypes. Nonetheless, due to its variability, influenza viruses can easily adapt to novel hosts, thereby creating antigenically novel subtypes [85].

Influenza virus A infects a broad spectrum of warm-blooded animals, including birds, and mammals, including horses, dogs, pigs, whales, minks and humans [85]. In birds, symptoms can vary from asymptomatic to mild respiratory infections, to fatal systemic disease, depending on virulence of the virus, host immune status, and bacterial co-infections [91].

Influenza viruses are typically diagnosed by virus isolation, ELISA, hemagglutination inhibition and neuraminidase inhibition tests [92, 93]. However, this method is retrospective and time-consuming, and faster and more sensitive methods, *i.e.*, molecular amplification methods are required for better outbreak containment [94]. Whereas PCR assays using degenerate oligonucleotides that are capable of detecting all HA or NA subtypes can provide high throughput for viral detection [95-97], expectedly, molecular surveillance of influenza viruses is highly dependent on the subtype of the virus. RT-PCR assays that specifically target HA, NA or matrix genes have been developed for subtyping of influenza viruses. For example, Lu *et al.* and Chen *et al.* designed the PCR that detects all H5 subtypes, while showing negativity to H1 subtypes [98, 99]. Furthermore, advances in PCR technology allow for simultaneous detection and differentiation of multiple influenza virus subtypes. While most of these methods use Taqman probes for fluorescence acquisition, SYBR green methods are also used, especially since the dissociation temperature of the amplification product can provide additional information [100]. On the other hand, Wu *et al.* use four sets of oligonucleotides to specifically detect influenza A and B, at the same time distinguishing subtypes H5N1 from other subtypes [101]. These methods could be very useful for large-scale screening of clinical samples for influenza virus in humans as well as poultry.

2.7. Paramyxoviruses

Paramyxoviruses are also Class V (negative sense single-stranded RNA) viruses. Members of this family include human respiratory syncytial virus and the human parainfluenza viruses, measles and mumps viruses, the zoonotic Hendra and Nipah viruses, canine distemper virus, Rinderpest virus and Newcastle disease virus [102]. Paramyxoviruses have been used to demonstrate how viral fusion proteins facilitate entry into host cells, and are potent inducers of Type 1 interferons [103-105].

Canine Distemper Virus. Canine Distemper Virus (CDV) is a species of the genus Morbili virus, which also includes the human measles virus. Domestic dogs are the most typical hosts, but the host spectrum of CDV also includes tigers, lions, leopards, foxes, ferrets, and minks, as well as marine mammals such as seals [106]. In the acute disease, CDV causes fever and leucopenia that accompany mucosal inflammation. The resulting symptoms include coughing and shivering, conjunctivitis, nasal discharge, pneumonia, diarrhea, and vomiting [107]. After the acute phase, CDV may invade epithelial tissues and the central nervous system. The resulting symptoms in the secondary disease phase are i) pustular dermatitis and hyperkeratosis (callusing) of nose and foot pads (hence "hard pad disease"), and ii) neurological disorders that include encephalitis associated with myoclonus, seizures, tremors, imbalance, ataxia, and limb weakness [106, 107]. The variability of signs makes clinical diagnosis relatively difficult. Myoclonus appears to be the only neurological sign highly suggestive of distemper infection. Laboratory detection methods in use such as virus neutralization assays, ELISA and nucleic acid hybridization assays are generally very time-consuming. Shimizu *et al.* first used PCR to detect morbilliviruses, including CDV, by targeting the PBGD gene (porphobilonigen deaminase), followed by Shin *et al.*, whose RT-PCR assay amplified the nucleoprotein gene [108, 109]. Various systems of CDV PCR have been developed and are widely employed, such as at the aforementioned Auburn University Molecular Diagnostics Laboratory, although the quality of sample source, nucleic acid extraction, and primer specificity is still critical [110].

Newcastle Disease Virus. Newcastle Disease Virus is a species of the genus Avulavirus. Also called avian paramyxovirus serotype 1 (APMV-1) virus, it causes Newcastle Disease in wild and domestic birds. Since its discovery in 1926, it has caused significant economic losses in the domestic poultry industry worldwide, especially in developing countries, due to its high susceptibility and the potentially severe diseases it produces [111, 112]. Interestingly, NDV has been used as a virutherapy agent in humans, since it was demonstrated that NDV has a potent oncolytic and immune-stimulatory effect [113, 114]. Symptoms of NDV can vary with virulence of the virus and the species, age and immune status of the host. They can range from respiratory and neurological signs to watery diarrhea, malformed eggs and reduced egg production, or sudden death in acute cases [115]. Clinical diagnosis of NDV is usually performed by virus isolation in embryonating chicken eggs (ECE), serology using the hemagglutination-inhibition (HI) test, or by RT-PCR. As early as 1991, PCR had been used for detection of NDV [116]. To account for the vast genetic variability of NDV, improved PCR assays have been developed that can detect all NDV strains, including the less virulent isolates [117]. The development of these assays has enabled clinicians, as well as researchers, to not only more effectively survey and control the disease, but also to better understand the evolution of NDV [111].

Rinderpest Virus. Rinderpest virus (RPV) is a member of the genus of Morbillivirus, together with canine distemper virus and the human measles virus [118]. It causes rinderpest (cattle plague), an ancient disease of nearly 100% mortality in cattle, Asian buffaloes, yaks, and many other artiodactyls. Rinderpest is characterized by fever, oral erosions, bloody diarrhea, and lymphoid necrosis before the animals die within 2 weeks of symptom onset [119, 120]. Throughout history since the Roman times, there have been periodic epizootics of rinderpest, sometimes killing all the cattle in the affected area [121]. However, the popularization of potent recombinant vaccines has successfully prevented the disease in the past few decades. On 14 October 2010, the Food and Agriculture Organization (FAO) of the United Nations announced that rinderpest had been completely eradicated, with no diagnoses for nine years [122, 123]. This makes RPV only the second viral disease to be eradicated, with the first being the smallpox virus in humans[121]. PCR diagnosis of RPV is relatively insignificant, since the virus is typically easy to diagnose, due to its severe symptoms and high mortality. However, wide usage of vaccines and emergence of less virulent strains made clinical confirmation more difficult. Before molecular techniques such as ELISA and PCR, RPV was confirmed by virus neutralization, agar gel immunodiffusion, virus isolation in cell culture, and sometimes by reproducing the disease in naïve animals [124]. PCR provides a highly sensitive platform for confirmation of rinderpest. For example, Forsyth used RT-PCR to differentially diagnose RPV from peste des petits virus (PPRV) [125], and, combining with a SNAP-ELISA, to differentiate the wild-type virus strain from vaccines [126].

2.8. Parvoviruses

Parvoviruses are Class II (positive sense single-stranded DNA) viruses. Members of the family Parvoviridae include canine minute virus and canine parvovirus in dogs, feline panleukopenia (also known as feline distemper) virus in cats, infectious hypodermal and hematopoietic necrosis (IHHN) in penaeid shrimp, and swine parvovirus in pigs [127]. Human parvovirus B19, another species of this group, causes erythema infectiosum, aplastic anemia and lethal cytopenias, particularly in children and immunocompromised patients [128, 129]. Like Newcastle disease virus, human parvovirus B19 has also been used in cancer treatment due to its oncolytic activities [130].

Canine parvovirus is a species of the genus Parvovirus. Two variants of canine parvovirus exist, canine parvovirus type 1 (CPV1, also called canine minute virus) and canine parvovirus type 2 (CPV2), with CPV2 causing more severe diseases. CPV2 can cause intestinal or cardiac diseases, especially in young puppies [131, 132]. CPV2 is typically diagnosed by detection of CPV2 in the feces, by EIA, by hemagglutination inhibition test, or by electron microscopy. Since PCR became available, it has become increasingly commonly used, due to its high sensitivity and specificity. Harasawa first used the VP1/VP2 genes as amplification targets to detect canine parvovirus [133], a method which was later improved so the assay could differentiate the wild type virus from vaccine strains [134].

2.9. Retroviruses

Retroviruses are Class VI (positive sense single-stranded RNA-RT) viruses. Many retroviruses are thought to be permanently integrated into the host genome [135]. For example, some 8% of human DNA represents fossil retroviral genomes [136]. These so-called endogenous retroviruses (ERVs) can participate in host genetic recombination events, and are implicated in autoimmunity and cancer [135, 137]. Retroviruses are found in most livestock and companion animals [138]. Important retroviruses include Abelson murine leukemia virus, bovine leukemia virus, equine infectious anemia virus, feline immunodeficiency virus (FIV), feline leukemia virus and simian retrovirus [139].

Feline Immunodeficiency Virus. FIV is a member of the genus Lentivirus that infects cats. As it is structurally and pathophysiologically related to the human immunodeficiency virus (HIV), it is used as a model for studies of lentivirus [140, 141]. Based on DNA sequences of the viral envelope and polymerase, five subtypes of FIV have been identified [141]. FIV infects about 2.5% of cats in the USA, but does not always cause fatal disease [142]. In cats, it can cause acute, subclinical or chronic symptoms, including fever, depression, generalized lymphadenopathy, stomatitis, odontoclasia, periodontitis, gingivitis, rhinitis, conjunctivitis, pneumonitis, enteritis, and dermatitis [143]. Classically, serum antibody titers against FIV are used to detect FIV. However, vaccinated cats also have positive titers, thereby creating false positive results. PCR can confirm FIV infection by directly detecting the presence/absence of the virus [144]. Hohdatsu *et al.* first demonstrated that PCR can be used for detection of proviral FIV DNA in peripheral blood lymphocytes [145]. Since then, improved assays have been designed to quantify the amount of FIV for better understanding of the viral replication status [146, 147]. Wang *et al.*'s real-time PCR is capable of not only differentiating different subtypes of FIV, but also distinguishing between naturally infected and vaccinated cats [16].

2.10. Other Viruses

Foot and Mouth Disease Virus. Foot and mouth disease virus (FMDV) is a species of the family Picornaviridae. It is the etiologic agent of foot and mouth disease (sometimes referred to as hoof and mouth disease), an infectious and sometimes fatal disease in domestic and wild bovids, including cattle, water buffalo, sheep, goats, antelope, deer, and bison. The virus causes a high fever for two or three days, followed by blisters inside the mouth and on the feet that may rupture and cause lameness. Foot and mouth disease is a devastating disease that is responsible for very large economic losses in the cattle industry [148]. As the presence of clinical signs alone is often inconclusive, confirmative diagnosis is usually carried out by cell culture isolation, complement fixation test, ELISA or PCR [149]. Laor *et al.* used a PCR assay to amplify a conserved region of the FMDV RNA polymerase gene, and were able to identify all subtypes of FMDV [150].

Infectious Bursal Disease Virus. Infectious bursal disease virus (IBDV) is a species of the family Birnaviridae. It causes Infectious Bursal Disease (or Gumboro's Disease), a highly contagious disease of young chickens, characterized by immunosuppression and mortality generally at 3 to 6 weeks of age. IBD is one of the most economically important diseases to the poultry industry worldwide because it increases susceptibility to other diseases and negatively interferes with effective vaccination [151]. IBDV is effectively diagnosed by molecular techniques such as ELISAs and PCRs [152]. RT-PCRs, especially real-time RT-PCRs, have been successful at detecting and subtyping IBDV [153, 154].

2.11. Zoonotic Viruses

Zoonoses are diseases of animals that can be directly or indirectly transmitted to humans. Many viruses that are generally considered human pathogens are in fact zoonotic agents. For example, West Nile virus mainly infects birds, but is known to infect humans, horses, dogs, cats, bats, and rabbits. Yellow Fever is a human disease, but can also infect other primates, whereas rabies is found in all mammals. Meanwhile, many animal viruses are also important from a human perspective. The emergence of the SARS virus in the human population, for instance, demonstrates that animals can carry pathogens which may become infectious to humans. Avian influenza viruses, on the other hand, can directly infect humans. Researches on human viruses and animal viruses are essentially inseparable; both make important contributions to our understanding of viruses in general, their replication, molecular biology, evolution and interaction with the host. Among the viruses that are responsible for the most significant viral zoonoses are influenza viruses, hepatitis E virus, SARS virus, foot and mouth disease virus (FMDV), arenaviruses, hantaviruses, and rabies virus. PCR detection of some viral zoonoses can be found in Table **1**.

2.12. Viruses not Typically Diagnosed by PCR

Prions. Prions are infectious agents composed of protein in a misfolded form. It is a distinct form of pathogen, in contrast to all others, which must contain nucleic acids, in the form of either DNA or RNA, along with protein components. Prions are responsible for the transmissible spongiform encephalopathies (TSEs) in a variety of mammals, most notably bovine spongiform encephalopathy (BSE, also known as "mad cow disease") in cattle, Scrapie in goats and sheep, and Creutzfeldt-Jakob disease (CJD) in humans [155, 156]. Because of the nature of prions, they cannot be directly diagnosed by nucleic acid detection methods, such as PCR, although it is possible to type the host susceptibility locus by PCR [157, 158]. Recently, Reuter *et al.* have developed a novel immune-PCR to quantitatively detect prions, using a direct conjugate of a prion-specific antibody (ICSM35) and a synthetic 99-bp DNA tail [11].

Poxviruses. Poxviruses are Class I (double-stranded DNA) viruses. There are two subfamilies of Poxviridae, Chordopoxvirinae and Entomopoxvirinae. The former are viruses that infect mostly vertebrates, whereas the latter infect insects [159]. Important members of the subfamily Chordopoxvirinae include fowlpox virus, pigeonpox virus, turkeypox virus, goatpox virus, sheeppox virus, rabbitpox virus, camelpox virus, cowpox virus, monkeypox virus, racoonpox virus, swinepox virus, squirrelpox virus, crocodilepox virus, *etc.* As the names indicate, poxviruses can infect a wide spectrum of species [160]. Variola virus, the causative agent of smallpox, one of the most devastating diseases known to humans but declared eradicated by the WHO in 1979, is also a species of Chordopoxvirinae [161]. Cowpox virus has also been identified as an emerging zoonosis recently [162, 163]. As with adenoviruses, poxviruses are pursed as recombinant vectors in cancer therapy [164]. Due to the distinct symptoms (cutaneous lesions) that poxviruses cause, most poxvirus infections can be recognized clinically. The virions can be recognized with negative staining and electron microscopy. Infections usually induce a humoral response that includes hemagglutination inhibition (HI), complement fixing (CF), and neutralizing antibodies, which can also be used for diagnosis. As poxvirus diseases are now relatively rare, PCR is not typically used for detection of poxviruses [165, 166].

Table 1: List of viral agents detected by PCR

Category	Virus	Host	Target Gene	References
Adenovirus	Avian adenovirus	Chicken	Hexon	[22, 25]
		Turkey	Hexon	[23]
	Canine adenovirus Type 1	Dog	E3	[34, 35]
Arenavirus	Lymphocytic choriomeningitis virus	Rodents	NP and GP genes	[167]
	Junin virus	Rodents	S RNA	[168]
Birnavirus	Infectious Bursal Disease virus	Chicken	Variable region of VP2 gene	[152]
			28S rRNA	[153]
Bunyavirus	Muerto Canyon virus	Rodents	M segment	[169]
Calicivirus	Feline calicivirus	Cat	Capsid	[42-44]
Coronavirus	Canine coronavirus	Dog	M gene	[56]
Coronavirus	Feline coronavirus	Cat	M gene	[64]
Filovirus	Ebola virus	Mammals	Polymerase	[170]

			L gene	[171]
Flavivirus	Bovine viral diarrhea virus	Cattle	p80 and gp53 (envelope glycoprotein)	[72]
			gp48	[73]
			5'-UTR	[8]
Herpesvirus	Madek's Disease virus	Chicken	132 bp tandem repeats	[79]
			Thymidine kinase	[80]
	Bovine herpesvirus 1 (BHV-1)	Cattle	Glycoprotein	[85, 86]
			dUTPase	[87]
	Bovine herpesvirus 2 (BHV-2)	Cattle	DNA polymerase (UL30)	[172]
Orthomyxovirus	Influenza virus A	Birds and mammals	HA_0 cleavage site sequence	[95]
			Haemagglutinin	[98]
			Neurominidase	[96]
Paramyxovirus	Canine distemper virus	Dog	Porphobilinogen deaminase (PBGD)	[108]
			NP (nucleocapsid protein)	[109]
			P and F genes	[173]
Paramyxovirus	Newcastle disease virus	Birds	Fusion protein F gene	[116]
			Matrix polymerase	[117]
Paramyxovirus	Rinderpest virus	Cattle	P and F genes	[125]
			Nucleocapsid	[126]
Parvovirus	Canine parvovirus	Dog	VP1/VP2	[133]
			VP2	[174]
Picornavirus	Foot and mouth disease virus	Cattle	RNA polymerase	[150]
			Capsid coding region	[175]
Poxvirus	Capripoxvirus	Sheep, goat and cattle	Attachment gene and fusion gene	[166]
Retrovirus	Feline immunodeficiency virus	Cat	*gag* gene	[16, 145]
			gag and *pol* gene	[176]
Rhabdovirus	Rabies virus	All mammals	N gene	[177, 178]

3. CONCLUSION

PCR is used for detection of many infectious agents because: i). it is fast and relatively easy to operate, ii). it provides high sensitivity and specificity, iii). it is a high-throughput method, so it can be used for large scale screening and typing. Additionally, if carefully designed, it can simultaneously detect non-infectious viruses and bacteria in clinical samples. PCR can be used for essentially all viruses, and is routinely used for a wide variety of viruses including swine fever virus, foot and mouth disease virus, Aujeszky's disease virus, porcine reproductive and respiratory syndrome virus, Newcastle disease virus, *etc.* With variations of PCR, veterinarians and researchers can now distinguish viral infections from bacterial infections, between different viral diseases, and identify multiple viral subtypes, often all in one step, with very high accuracy and reproducibility. Because of the high sensitivity, however, PCR assays should be carried out by appropriately trained personnel, and with extreme caution, to avoid contaminations, while maximizing efficiency and sensitivity.

REFERENCES

[1]   Le BM, Presti R. The current state of viral diagnostics for respiratory infections. Mo Med 2009; 106: 283-6.
[2]   Ratcliff RM, Chang G, Kok T, Sloots TP. Molecular diagnosis of medical viruses. Curr Issues Mol Biol 2007; 9: 87-102.
[3]   Belak S, Thoren P. Molecular diagnosis of animal diseases: some experiences over the past decade. Expert Rev Mol Diagn 2001; 1: 434-43.

[4]   Locher F, Suryanarayana VV, Tratschin JD. Rapid detection and characterization of foot-and-mouth disease virus by restriction enzyme and nucleotide sequence analysis of PCR products. J Clin Microbiol 1995; 33: 440-4.

[5]   Rodriguez A, Nunez JI, Nolasco G, Ponz F, Sobrino F, de Blas C. Direct PCR detection of foot-and-mouth disease virus. J Virol Methods 1994; 47: 345-9.

[6]   Chomczynski P, Sacchi N. Single-step method of RNA isolation by acid guanidinium thiocyanate-phenol-chloroform extraction. Anal Biochem 1987; 162: 156-9.

[7]   Belak S. Molecular diagnosis of viral diseases, present trends and future aspects A view from the OIE Collaborating Centre for the Application of Polymerase Chain Reaction Methods for Diagnosis of Viral Diseases in Veterinary Medicine. Vaccine 2007; 25: 5444-52.

[8]   Baxi M, McRae D, Baxi S, *et al.* A one-step multiplex real-time RT-PCR for detection and typing of bovine viral diarrhea viruses. Vet Microbiol 2006; 116: 37-44.

[9]   Caterina KM, Frasca S, Jr., Girshick T, Khan MI. Development of a multiplex PCR for detection of avian adenovirus, avian reovirus, infectious bursal disease virus, and chicken anemia virus. Mol Cell Probes 2004; 18: 293-8.

[10]   Sykes JE, Allen JL, Studdert VP, Browning GF. Detection of feline calicivirus, feline herpesvirus 1 and Chlamydia psittaci mucosal swabs by multiplex RT-PCR/PCR. Vet Microbiol 2001; 81: 95-108.

[11]   Reuter T, Gilroyed BH, Alexander TW, *et al.* Prion protein detection *via* direct immuno-quantitative real-time PCR. J Microbiol Methods 2009; 78: 307-11.

[12]   Ohe K, Sakai S, Sunaga F, Murakami M, *et al.* Detection of feline calicivirus (FCV) from vaccinated cats and phylogenetic analysis of its capsid genes. Vet Res Commun 2006; 30: 293-305.

[13]   Jarmin S, Manvell R, Gough RE, Laidlaw SM, Skinner MA. Avipoxvirus phylogenetics: identification of a PCR length polymorphism that discriminates between the two major clades. J Gen Virol 2006; 87: 2191-201.

[14]   Ottiger HP. Development, standardization and assessment of PCR systems for purity testing of avian viral vaccines. Biologicals 2010; 38: 381-8.

[15]   Nuanualsuwan S, Cliver DO. Pretreatment to avoid positive RT-PCR results with inactivated viruses. J Virol Methods 2002; 104: 217-25.

[16]   Wang C, Johnson CM, Ahluwalia SK, *et al.* Dual-emission fluorescence resonance energy transfer (FRET) real-time PCR differentiates feline immunodeficiency virus subtypes and discriminates infected from vaccinated cats. J Clin Microbiol 2010; 48: 1667-72.

[17]   Harrison SC. Virology. Looking inside adenovirus. Science 2010; 329: 1026-7.

[18]   Katoh H, Ogawa H, Ohya K, Fukushi H. A review of DNA viral infections in psittacine birds. J Vet Med Sci 2010; 72: 1099-106.

[19]   McFerran JB, Adair BM. Avian adenoviruses-a review. Avian Pathol 1977; 6: 189-217.

[20]   McFerran JB, Smyth JA. Avian adenoviruses. Rev Sci Tech 2000; 19: 589-601.

[21]   Hess M. Detection and differentiation of avian adenoviruses: a review. Avian Pathol 2000; 29: 195-206.

[22]   Raue R, Hess M. Hexon based PCRs combined with restriction enzyme analysis for rapid detection and differentiation of fowl adenoviruses and egg drop syndrome virus. J Virol Methods 1998; 73: 211-7.

[23]   Hess M, Raue R, Hafez HM. PCR for specific detection of haemorrhagic enteritis virus of turkeys, an avian adenovirus. J Virol Methods 1999; 81: 199-203.

[24]   Mase M, Mitake H, Inoue T, Imada T. Identification of group I-III avian adenovirus by PCR coupled with direct sequencing of the hexon gene. J Vet Med Sci 2009; 71: 1239-42.

[25]   Ganesh K, Suryanarayana VV, Raghavan R. Detection of fowl adenovirus associated with hydropericardium hepatitis syndrome by a polymerase chain reaction. Vet Res Commun 2002; 26: 73-80.

[26]   Romanova N, Corredor JC, Nagy E. Detection and quantitation of fowl adenovirus genome by a real-time PCR assay. J Virol Methods 2009; 159: 58-63.

[27]   Katoh H, Ohya K, Fukushi H. Development of novel real-time PCR assays for detecting DNA virus infections in psittaciform birds. J Virol Methods 2008; 154: 92-8.

[28]   Luschow D, Prusas C, Lierz M, Gerlach H, Soike D, Hafez HM. Adenovirus of psittacine birds: investigations on isolation and development of a real-time polymerase chain reaction for specific detection. Avian Pathol 2007; 36: 487-94.

[29]   Decaro N, Martella V, Buonavoglia C. Canine adenoviruses and herpesvirus. Vet Clin North Am Small Anim Pract 2008; 38: 799-814, viii.

[30]   Watson PJ. Chronic hepatitis in dogs: a review of current understanding of the aetiology, progression, and treatment. Vet J 2004; 167: 228-41.

[31] Boomkens SY, Penning LC, Egberink HF, van den Ingh TS, Rothuizen J. Hepatitis with special reference to dogs. A review on the pathogenesis and infectious etiologies, including unpublished results of recent own studies. Vet Q 2004; 26: 107-14.

[32] Noon KF, Rogul M, Binn LN, *et al.* An enzyme-linked immunosorbent assay for the detection of canine antibodies to canine adenoviruses. Lab Anim Sci 1979; 29: 603-9.

[33] Kiss I, Matiz K, Bajmoci E, Rusvai M, Harrach B. Infectious canine hepatitis: detection of canine adenovirus type 1 by polymerase chain reaction. Acta Vet Hung 1996; 44: 253-8.

[34] Hu RL, Huang G, Qiu W, Zhong ZH, Xia XZ, Yin Z. Detection and differentiation of CAV-1 and CAV-2 by polymerase chain reaction. Vet Res Commun 2001; 25: 77-84.

[35] Chaturvedi U, Tiwari AK, Ratta B, *et al.* Detection of canine adenoviral infections in urine and faeces by the polymerase chain reaction. J Virol Methods 2008; 149: 260-3.

[36] Boomkens SY, Slump E, Egberink HF, Rothuizen J, Penning LC. PCR screening for candidate etiological agents of canine hepatitis. Vet Microbiol 2005; 108: 49-55.

[37] Fenner F. Veterinary virology. Orlando: Academic Press; 1987.

[38] Hurley KF, Sykes JE. Update on feline calicivirus: new trends. Vet Clin North Am Small Anim Pract 2003; 33: 759-72.

[39] Radford AD, Coyne KP, Dawson S, Porter CJ, Gaskell RM. Feline calicivirus. Vet Res 2007; 38: 319-35.

[40] Pesavento PA, Chang KO, Parker JS. Molecular virology of feline calicivirus. Vet Clin North Am Small Anim Pract 2008; 38: 775-86, vii.

[41] Thiel HJ, Konig M. Caliciviruses: an overview. Vet Microbiol 1999; 69: 55-62.

[42] Radford AD, Bennett M, McArdle F, *et al.* The use of sequence analysis of a feline calicivirus (FCV) hypervariable region in the epidemiological investigation of FCV related disease and vaccine failures. Vaccine 1997; 15: 1451-8.

[43] Kreutz LC, Johnson RP, Seal BS. Phenotypic and genotypic variation of feline calicivirus during persistent infection of cats. Vet Microbiol 1998; 59: 229-36.

[44] Sykes JE, Studdert VP, Browning GF. Detection and strain differentiation of feline calicivirus in conjunctival swabs by RT-PCR of the hypervariable region of the capsid protein gene. Arch Virol 1998; 143: 1321-34.

[45] Radford AD, Dawson S, Wharmby C, Ryvar R, Gaskell RM. Comparison of serological and sequence-based methods for typing feline calcivirus isolates from vaccine failures. Vet Rec 2000; 146: 117-23.

[46] Cavanagh D. Coronavirus avian infectious bronchitis virus. Vet Res 2007; 38: 281-97.

[47] Ignjatovic J, Sapats S. Avian infectious bronchitis virus. Rev Sci Tech 2000; 19: 493-508.

[48] Schwegmann-Wessels C, Herrler G. Transmissible gastroenteritis virus infection: a vanishing specter. Dtsch Tierarztl Wochenschr 2006; 113: 157-9.

[49] Saif LJ, van Cott JL, Brim TA. Immunity to transmissible gastroenteritis virus and porcine respiratory coronavirus infections in swine. Vet Immunol Immunopathol 1994; 43: 89-97.

[50] Laude H, Van Reeth K, Pensaert M. Porcine respiratory coronavirus: molecular features and virus-host interactions. Vet Res 1993; 24: 125-50.

[51] Decaro N, Buonavoglia C. An update on canine coronaviruses: viral evolution and pathobiology. Vet Microbiol 2008; 132: 221-34.

[52] Schulz BS, Strauch C, Mueller RS, Eichhorn W, Hartmann K. Comparison of the prevalence of enteric viruses in healthy dogs and those with acute haemorrhagic diarrhoea by electron microscopy. J Small Anim Pract 2008; 49: 84-8.

[53] Erles K, Toomey C, Brooks HW, Brownlie J. Detection of a group 2 coronavirus in dogs with canine infectious respiratory disease. Virology 2003; 310: 216-23.

[54] Erles K, Brownlie J. Investigation into the causes of canine infectious respiratory disease: antibody responses to canine respiratory coronavirus and canine herpesvirus in two kennelled dog populations. Arch Virol 2005; 150: 1493-504.

[55] Erles K, Shiu KB, Brownlie J. Isolation and sequence analysis of canine respiratory coronavirus. Virus Res 2007; 124: 78-87.

[56] Decaro N, Martella V, Ricci D, *et al.* Genotype-specific fluorogenic RT-PCR assays for the detection and quantitation of canine coronavirus type I and type II RNA in faecal samples of dogs. J Virol Methods 2005; 130: 72-8.

[57] Buonavoglia C, Decaro N, Martella V, *et al.* Canine coronavirus highly pathogenic for dogs. Emerg Infect Dis 2006; 12: 492-4.

[58]   Pedersen NC. A review of feline infectious peritonitis virus infection: 1963-2008. J Feline Med Surg 2009; 11: 225-58.

[59]   Paltrinieri S, Grieco V, Comazzi S, Cammarata Parodi M. Laboratory profiles in cats with different pathological and immunohistochemical findings due to feline infectious peritonitis (FIP). J Feline Med Surg 2001; 3: 149-59.

[60]   Addie D, Belak S, Boucraut-Baralon C, *et al.* Feline infectious peritonitis. ABCD guidelines on prevention and management. J Feline Med Surg 2009; 11: 594-604.

[61]   Andrew SE. Feline infectious peritonitis. Vet Clin North Am Small Anim Pract 2000; 30: 987-1000.

[62]   Nghiem PP, Schatzberg SJ. Conventional and molecular diagnostic testing for the acute neurologic patient. J Vet Emerg Crit Care (San Antonio) 2010; 20: 46-61.

[63]   Sharif S, Arshad SS, Hair-Bejo M, Omar AR, Zeenathul NA, Alazawy A. Diagnostic methods for feline coronavirus: a review. Vet Med Int 2010; 2010.

[64]   Simons FA, Vennema H, Rofina JE, *et al.* A mRNA PCR for the diagnosis of feline infectious peritonitis. J Virol Methods 2005; 124: 111-6.

[65]   Neyts J, Leyssen P, De Clercq E. Infections with flaviviridae. Verh K Acad Geneeskd Belg 1999; 61: 661-97; discussion 97-9.

[66]   Houe H. Economic impact of BVDV infection in dairies. Biologicals 2003; 31: 137-43.

[67]   Fulton RW. Bovine respiratory disease research (1983-2009). Anim Health Res Rev 2009; 10: 131-9.

[68]   Kelling CL. Evolution of bovine viral diarrhea virus vaccines. Vet Clin North Am Food Anim Pract 2004; 20: 115-29.

[69]   Moennig V, Houe H, Lindberg A. BVD control in Europe: current status and perspectives. Anim Health Res Rev 2005; 6: 63-74.

[70]   Brock KV. Strategies for the control and prevention of bovine viral diarrhea virus. Vet Clin North Am Food Anim Pract 2004; 20: 171-80.

[71]   Brodersen BW. Immunohistochemistry used as a screening method for persistent bovine viral diarrhea virus infection. Vet Clin North Am Food Anim Pract 2004; 20: 85-93.

[72]   Hertig C, Pauli U, Zanoni R, Peterhans E. Detection of bovine viral diarrhea (BVD) virus using the polymerase chain reaction. Vet Microbiol 1991; 26: 65-76.

[73]   Belak S, Ballagi-Pordany A. Bovine viral diarrhea virus infection: rapid diagnosis by the polymerase chain reaction. Arch Virol Suppl 1991; 3: 181-90.

[74]   Letellier C, Kerkhofs P. Real-time PCR for simultaneous detection and genotyping of bovine viral diarrhea virus. J Virol Methods 2003; 114: 21-7.

[75]   Fabricant CG, Fabricant J. Atherosclerosis induced by infection with Marek's disease herpesvirus in chickens. Am Heart J 1999; 138: S465-8.

[76]   Baigent SJ, Smith LP, Nair VK, Currie RJ. Vaccinal control of Marek's disease: current challenges, and future strategies to maximize protection. Vet Immunol Immunopathol 2006; 112: 78-86.

[77]   Islam AF, Wong CW, Walkden-Brown SW, Colditz IG, Arzey KE, Groves PJ. Immunosuppressive effects of Marek's disease virus (MDV) and herpesvirus of turkeys (HVT) in broiler chickens and the protective effect of HVT vaccination against MDV challenge. Avian Pathol 2002; 31: 449-61.

[78]   Gimeno IM. Marek's disease vaccines: a solution for today but a worry for tomorrow? Vaccine 2008; 26 Suppl 3: C31-41.

[79]   Silva RF. Differentiation of pathogenic and non-pathogenic serotype 1 Marek's disease viruses (MDVs) by the polymerase chain reaction amplification of the tandem direct repeats within the MDV genome. Avian Dis 1992; 36: 521-8.

[80]   Rong-Fu W, Beasley JN, Cao WW, Slavik MF, Johnson MG. Development of PCR method specific for Marek's disease virus. Mol Cell Probes 1993; 7: 127-31.

[81]   Reimann I, Werner O. Use of the polymerase chain reaction for the detection of reticuloendotheliosis virus in Marek's disease vaccines and chicken tissues. Zentralbl Veterinarmed B 1996; 43: 75-84.

[82]   Jones C, Chowdhury S. Bovine herpesvirus type 1 (BHV-1) is an important cofactor in the bovine respiratory disease complex. Vet Clin North Am Food Anim Pract 2010; 26: 303-21.

[83]   Shah AP, Youngquist ST, McClung CD, *et al.* Markers of Progenitor Cell Recruitment and Differentiation Rise Early During Ischemia and Continue During Resuscitation in a Porcine Acute Ischemia Model. J Interferon Cytokine Res 2011.

[84]   Jotzu C, Alt E, Welte G, Li J, *et al.* Adipose tissue derived stem cells differentiate into carcinoma-associated fibroblast-like cells under the influence of tumor derived factors. Cell Oncol (Dordr) 2011.

[85]   Singla DK, Singla RD, Lamm S, Glass C. TGF{beta}2 Treatment Enhances Cytoprotective Factors Released from Embryonic Stem Cells and Inhibits Apoptosis in the Infarcted Myocardium. Am J Physiol Heart Circ Physiol 2011.

[86]   Gotte M, Greve B, Kelsch R, *et al.* The adult stem cell marker musashi-1 modulates endometrial carcinoma cell cycle progression and apoptosis *via* notch-1 and p21(WAF1/CIP1). Int J Cancer 2010.

[87]   Su JZ, Cai WQ, Lin MP, Lin MH, Lin M. [Transplantation of mesenchymal stem cells transfected with hepatocyte growth factor gene improves heart function in rats with heart failure]. Zhongguo Ying Yong Sheng Li Xue Za Zhi 2009; 25: 521-6.

[88]   Wang Y, Abarbanell AM, Herrmann JL, *et al.* TLR4 inhibits mesenchymal stem cell (MSC) STAT3 activation and thereby exerts deleterious effects on MSC-mediated cardioprotection. PLoS One 2010; 5: e14206.

[89]   Manferdini C, Gabusi E, Grassi F, *et al.* Evidence of specific characteristics and osteogenic potentiality in bone cells from Tibia. J Cell Physiol 2011.

[90]   Yassine HM, Lee CW, Gourapura R, Saif YM. Interspecies and intraspecies transmission of influenza A viruses: viral, host and environmental factors. Anim Health Res Rev 2010; 11: 53-72.

[91]   Mohammadi R, Azizi S, Delirezh N, Hobbenaghi R, Amini K. Comparison of beneficial effects of undifferentiated cultured bone marrow stromal cells and omental adipose-derived nucleated cell fractions on sciatic nerve regeneration. Muscle Nerve 2011; 43: 157-63.

[92]   Bokma BH, Hall C, Siegfried LM, Weaver JT. Surveillance for avian influenza in the United States. Ann N Y Acad Sci 2006; 1081: 163-8.

[93]   Charlton B, Crossley B, Hietala S. Conventional and future diagnostics for avian influenza. Comp Immunol Microbiol Infect Dis 2009; 32: 341-50.

[94]   Wang R, Taubenberger JK. Methods for molecular surveillance of influenza. Expert Rev Anti Infect Ther 2010; 8: 517-27.

[95]   Gall A, Hoffmann B, Harder T, Grund C, Beer M. Universal primer set for amplification and sequencing of HA0 cleavage sites of all influenza A viruses. J Clin Microbiol 2008; 46: 2561-7.

[96]   Alvarez AC, Brunck ME, Boyd V, *et al.* A broad spectrum, one-step reverse-transcription PCR amplification of the neuraminidase gene from multiple subtypes of influenza A virus. Virol J 2008; 5: 77.

[97]   Pasick J. Advances in the molecular based techniques for the diagnosis and characterization of avian influenza virus infections. Transbound Emerg Dis 2008; 55: 329-38.

[98]   Lu YY, Yan JY, Feng Y, Xu CP, Shi W, Mao HY. Rapid detection of H5 avian influenza virus by TaqMan-MGB real-time RT-PCR. Lett Appl Microbiol 2008; 46: 20-5.

[99]   Chen W, He B, Li C, *et al.* Real-time RT-PCR for H5N1 avian influenza A virus detection. J Med Microbiol 2007; 56: 603-7.

[100]   Krafft AE, Russell KL, Hawksworth AW, *et al.* Evaluation of PCR testing of ethanol-fixed nasal swab specimens as an augmented surveillance strategy for influenza virus and adenovirus identification. J Clin Microbiol 2005; 43: 1768-75.

[101]   Wu C, Cheng X, He J, Lv X, *et al.* A multiplex real-time RT-PCR for detection and identification of influenza virus types A and B and subtypes H5 and N1. J Virol Methods 2008; 148: 81-8.

[102]   Harrison MS, Sakaguchi T, Schmitt AP. Paramyxovirus assembly and budding: building particles that transmit infections. Int J Biochem Cell Biol 2010; 42: 1416-29.

[103]   Lamb RA, Jardetzky TS. Structural basis of viral invasion: lessons from paramyxovirus F. Curr Opin Struct Biol 2007; 17: 427-36.

[104]   Russell CJ, Luque LE. The structural basis of paramyxovirus invasion. Trends Microbiol 2006; 14: 243-6.

[105]   Goodbourn S, Randall RE. The regulation of type I interferon production by paramyxoviruses. J Interferon Cytokine Res 2009; 29: 539-47.

[106]   Vandevelde M, Zurbriggen A. Demyelination in canine distemper virus infection: a review. Acta Neuropathol 2005; 109: 56-68.

[107]   Martella V, Elia G, Buonavoglia C. Canine distemper virus. Vet Clin North Am Small Anim Pract 2008; 38: 787-97, vii-viii.

[108]   Shimizu H, Shimizu C, Burns JC. Detection of novel RNA viruses: morbilliviruses as a model system. Mol Cell Probes 1994; 8: 209-14.

[109]   Shin Y, Mori T, Okita M, Gemma T, Kai C, Mikami T. Detection of canine distemper virus nucleocapsid protein gene in canine peripheral blood mononuclear cells by RT-PCR. J Vet Med Sci 1995; 57: 439-45.

[110]   Elia G, Decaro N, Martella V, *et al.* Detection of canine distemper virus in dogs by real-time RT-PCR. J Virol Methods 2006; 136: 171-6.

[111]   Miller PJ, Decanini EL, Afonso CL. Newcastle disease: evolution of genotypes and the related diagnostic challenges. Infect Genet Evol 2010; 10: 26-35.

[112]   Swayne DE, King DJ. Avian influenza and Newcastle disease. J Am Vet Med Assoc 2003; 222: 1534-40.

[113]   Ravindra PV, Tiwari AK, Sharma B, Chauhan RS. Newcastle disease virus as an oncolytic agent. Indian J Med Res 2009; 130: 507-13.

[114]   Schirrmacher V, Fournier P. Newcastle disease virus: a promising vector for viral therapy, immune therapy, and gene therapy of cancer. Methods Mol Biol 2009; 542: 565-605.

[115]   Senne DA, King DJ, Kapczynski DR. Control of Newcastle disease by vaccination. Dev Biol (Basel) 2004; 119: 165-70.

[116]   Jestin V, Jestin A. Detection of Newcastle disease virus RNA in infected allantoic fluids by *in vitro* enzymatic amplification (PCR). Arch Virol 1991; 118: 151-61.

[117]   Mia Kim L, Suarez DL, Afonso CL. Detection of a broad range of class I and II Newcastle disease viruses using a multiplex real-time reverse transcription polymerase chain reaction assay. J Vet Diagn Invest 2008; 20: 414-25.

[118]   Haas L, Barrett T. Rinderpest and other animal morbillivirus infections: comparative aspects and recent developments. Zentralbl Veterinarmed B 1996; 43: 411-20.

[119]   Huygelen C. The immunization of cattle against rinderpest in eighteenth-century Europe. Med Hist 1997; 41: 182-96.

[120]   Roeder PL, Taylor WP. Rinderpest. Vet Clin North Am Food Anim Pract 2002; 18: 515-47, ix.

[121]   Crowther JR. Rinderpest: at war with the disease of war. Sci Prog 1997; 80 (Pt 1): 21-43.

[122]   Normile D. Rinderpest. Driven to extinction. Science 2008; 319: 1606-9.

[123]   Barrett T, Rossiter PB. Rinderpest: the disease and its impact on humans and animals. Adv Virus Res 1999; 53: 89-110.

[124]   Diallo A, Libeau G, Couacy-Hymann E, Barbron M. Recent developments in the diagnosis of rinderpest and peste des petits ruminants. Vet Microbiol 1995; 44: 307-17.

[125]   Forsyth MA, Barrett T. Evaluation of polymerase chain reaction for the detection and characterisation of rinderpest and peste des petits ruminants viruses for epidemiological studies. Virus Res 1995; 39: 151-63.

[126]   Forsyth MA, Parida S, Alexandersen S, Belsham GJ, Barrett T. Rinderpest virus lineage differentiation using RT-PCR and SNAP-ELISA. J Virol Methods 2003; 107: 29-36.

[127]   Lamm CG, Rezabek GB. Parvovirus infection in domestic companion animals. Vet Clin North Am Small Anim Pract 2008; 38: 837-50, viii-ix.

[128]   Human parvovirus B19. Am J Transplant 2004; 4 Suppl 10: 92-4.

[129]   Lunardi C, Tinazzi E, Bason C, Dolcino M, Corrocher R, Puccetti A. Human parvovirus B19 infection and autoimmunity. Autoimmun Rev 2008; 8: 116-20.

[130]   Rommelaere J, Geletneky K, Angelova AL, *et al.* Oncolytic parvoviruses as cancer therapeutics. Cytokine Growth Factor Rev 2010; 21: 185-95.

[131]   Goddard A, Leisewitz AL. Canine parvovirus. Vet Clin North Am Small Anim Pract 2010; 40: 1041-53.

[132]   Pollock RV, Coyne MJ. Canine parvovirus. Vet Clin North Am Small Anim Pract 1993; 23: 555-68.

[133]   Harasawa R, Yoshida M, Kataoka Y, Katae H. [Detection by PCR of genomic markers in canine parvovirus and feline panleukopenia virus]. C R Seances Soc Biol Fil 1993; 187: 554-60.

[134]   Hirasawa T, Yono K, Mikazuki K. Differentiation of wild- and vaccine-type canine parvoviruses by PCR and restriction-enzyme analysis. Zentralbl Veterinarmed B 1995; 42: 601-10.

[135]   Weiss RA. The discovery of endogenous retroviruses. Retrovirology 2006; 3: 67.

[136]   Griffiths DJ. Endogenous retroviruses in the human genome sequence. Genome Biol 2001; 2: REVIEWS1017.

[137]   Ruprecht K, Mayer J, Sauter M, Roemer K, Mueller-Lantzsch N. Endogenous retroviruses and cancer. Cell Mol Life Sci 2008; 65: 3366-82.

[138]   Miyazawa T. [Receptors for animal retroviruses]. Uirusu 2009; 59: 223-42.

[139]   Lerche NW. Simian retroviruses: infection and disease--implications for immunotoxicology research in primates. J Immunotoxicol 2010; 7: 93-101.

[140]   Elder JH, Lin YC, Fink E, Grant CK. Feline immunodeficiency virus (FIV) as a model for study of lentivirus infections: parallels with HIV. Curr HIV Res 2010; 8: 73-80.

[141]   Hayward JJ, Rodrigo AG. Molecular epidemiology of feline immunodeficiency virus in the domestic cat (Felis catus). Vet Immunol Immunopathol 2010; 134: 68-74.

[142]   Zislin A. Feline immunodeficiency virus vaccine: a rational paradigm for clinical decision-making. Biologicals 2005; 33: 219-20.

[143] Sykes JE. Immunodeficiencies caused by infectious diseases. Vet Clin North Am Small Anim Pract 2010; 40: 409-23.

[144] Hosie MJ, Addie D, Belak S, *et al.* Feline immunodeficiency. ABCD guidelines on prevention and management. J Feline Med Surg 2009; 11: 575-84.

[145] Hohdatsu T, Yamada M, Okada M, *et al.* Detection of feline immunodeficiency proviral DNA in peripheral blood lymphocytes by the polymerase chain reaction. Vet Microbiol 1992; 30: 113-23.

[146] Lawson M, Meers J, Blechynden L, Robinson W, Greene W, Carnegie P. The detection and quantification of feline immunodeficiency provirus in peripheral blood mononuclear cells using the polymerase chain reaction. Vet Microbiol 1993; 38: 11-21.

[147] Tomonaga K, Mikami T. Detection of feline immunodeficiency virus transcripts by quantitative reverse transcription-polymerase chain reaction. Vet Microbiol 1996; 48: 337-44.

[148] Rodriguez LL, Grubman MJ. Foot and mouth disease virus vaccines. Vaccine 2009; 27 Suppl 4: D90-4.

[149] Remond M, Kaiser C, Lebreton F. Diagnosis and screening of foot-and-mouth disease. Comp Immunol Microbiol Infect Dis 2002; 25: 309-20.

[150] Laor O, Torgersen H, Yadin H, Becker Y. Detection of FMDV RNA amplified by the polymerase chain reaction (PCR). J Virol Methods 1992; 36: 197-207.

[151] van den Berg TP, Eterradossi N, Toquin D, Meulemans G. Infectious bursal disease (Gumboro disease). Rev Sci Tech 2000; 19: 509-43.

[152] Jackwood DJ. Recent trends in the molecular diagnosis of infectious bursal disease viruses. Anim Health Res Rev 2004; 5: 313-6.

[153] Moody A, Sellers S, Bumstead N. Measuring infectious bursal disease virus RNA in blood by multiplex real-time quantitative RT-PCR. J Virol Methods 2000; 85: 55-64.

[154] Jackwood DJ, Sommer SE. Identification of infectious bursal disease virus quasispecies in commercial vaccines and field isolates of this double-stranded RNA virus. Virology 2002; 304: 105-13.

[155] Spero M, Lazibat I. Creutzfeldt-Jakob disease: case report and review of the literature. Acta Clin Croat 2010; 49: 181-7.

[156] Harman JL, Silva CJ. Bovine spongiform encephalopathy. J Am Vet Med Assoc 2009; 234: 59-72.

[157] Calero O, Hortiguela R, Albo C, de Pedro-Cuesta J, Calero M. Allelic discrimination of genetic human prion diseases by real-time PCR genotyping. Prion 2009; 3: 146-50.

[158] L'Homme Y, Leboeuf A, Cameron J. PrP genotype frequencies of Quebec sheep breeds determined by real-time PCR and molecular beacons. Can J Vet Res 2008; 72: 320-4.

[159] Hughes AL, Irausquin S, Friedman R. The evolutionary biology of poxviruses. Infect Genet Evol 2010; 10: 50-9.

[160] Werden SJ, Rahman MM, McFadden G. Poxvirus host range genes. Adv Virus Res 2008; 71: 135-71.

[161] Frenk J. Disease control priorities in developing countries. Salud Publica Mexico 2006; 48: 522-5.

[162] Vorou RM, Papavassiliou VG, Pierroutsakos IN. Cowpox virus infection: an emerging health threat. Curr Opin Infect Dis 2008; 21: 153-6.

[163] Essbauer S, Pfeffer M, Meyer H. Zoonotic poxviruses. Vet Microbiol 2010; 140: 229-36.

[164] Gilbert PA, McFadden G. Poxvirus cancer therapy. Recent Pat Antiinfect Drug Discov 2006; 1: 309-21.

[165] Markoulatos P, Mangana-Vougiouka O, Koptopoulos G, Nomikou K, Papadopoulos O. Detection of sheep poxvirus in skin biopsy samples by a multiplex polymerase chain reaction. J Virol Methods 2000; 84: 161-7.

[166] Ireland DC, Binepal YS. Improved detection of capripoxvirus in biopsy samples by PCR. J Virol Methods 1998; 74: 1-7.

[167] McCausland MM, Crotty S. Quantitative PCR technique for detecting lymphocytic choriomeningitis virus *in vivo*. J Virol Methods 2008; 147: 167-76.

[168] Lozano ME, Enria D, Maiztegui JI, Grau O, Romanowski V. Rapid diagnosis of Argentine hemorrhagic fever by reverse transcriptase PCR-based assay. J Clin Microbiol 1995; 33: 1327-32.

[169] Hjelle B, Spiropoulou CF, Torrez-Martinez N, Morzunov S, Peters CJ, Nichol ST. Detection of Muerto Canyon virus RNA in peripheral blood mononuclear cells from patients with hantavirus pulmonary syndrome. J Infect Dis 1994; 170: 1013-7.

[170] Onyango CO, Opoka ML, Ksiazek TG, *et al.* Laboratory diagnosis of Ebola hemorrhagic fever during an outbreak in Yambio, Sudan, 2004. J Infect Dis 2007; 196 Suppl 2: S193-8.

[171] Formenty P, Leroy EM, Epelboin A, *et al.* Detection of Ebola virus in oral fluid specimens during outbreaks of Ebola virus hemorrhagic fever in the Republic of Congo. Clin Infect Dis 2006; 42: 1521-6.

[172] Nishishita N, Ijiri H, Takenaka C, *et al.* The use of leukemia inhibitory factor immobilized on virus-derived polyhedra to support the proliferation of mouse embryonic and induced pluripotent stem cells. Biomaterials 2011.

[173] Goller KV, Fyumagwa RD, Nikolin V, *et al.* Fatal canine distemper infection in a pack of African wild dogs in the Serengeti ecosystem, Tanzania. Vet Microbiol 2010; 146: 245-52.

[174] Decaro N, Elia G, Martella V, *et al.* A real-time PCR assay for rapid detection and quantitation of canine parvovirus type 2 in the feces of dogs. Vet Microbiol 2005; 105: 19-28.

[175] Hofner MC, Carpenter WC, Donaldson AI. Detection of foot-and-mouth disease virus RNA in clinical samples and cell culture isolates by amplification of the capsid coding region. J Virol Methods 1993; 42: 53-61.

[176] Rimstad E, Ueland K. Detection of feline immunodeficiency virus by a nested polymerase chain reaction. J Virol Methods 1992; 36: 239-48.

[177] Crepin P, Audry L, Rotivel Y, Gacoin A, Caroff C, Bourhy H. Intravitam diagnosis of human rabies by PCR using saliva and cerebrospinal fluid. J Clin Microbiol 1998; 36: 1117-21.

[178] Wacharapluesadee S, Hemachudha T. Nucleic-acid sequence based amplification in the rapid diagnosis of rabies. Lancet 2001; 358: 892-3.

CHAPTER 6

Quantitative PCR as a Diagnostic Technique in Veterinary Parasitology

Hongzhuan Wu[1,*], Kirsten Jaegersen[2], Boakai K. Robertson[1], Robert Villafane[1] and Chengming Wang[3]

[1]*Department of Biological Sciences, Alabama State University, Montgomery, AL, 36101, USA,* [2]*Ross University School of Veterinary Medicine, P.O. Box 334, Basseterre, St. Kitts, West Indies, 00265 and* [3]*School of Veterinary Medicine, Yangzhou University, Jiangsu, China, 225009*

Abstract: Animals are routinely exposed to parasites from different taxonomic groups resulting in significant morbidity, mortality and economic losses. Accurate identification of the responsible parasites is central to the understanding and management of these infections and associated diseases. Comprehensive approaches to facilitate the diagnosis of parasites and parasitic diseases will yield better insight into their basic biology, epidemiology, pathogenicity, and the development of treatment strategies. Traditionally, the diagnosis of parasitic infections mainly relies on testing for the presence of parasites through direct fecal examination, blood smears, *etc,* but clinically, it is often difficult to elucidate the entire offending organism. Techniques for diagnosis of parasites such as counting parasites are often time-consuming, difficult, inaccurate, of limited sensitivity, and occasionally unpleasant. While the majority of parasites exhibit multiple stages during the course of their life cycles, the nucleic acids extracted from them during these different periods remain identical. PCR, providing exquisite sensitivity and specificity for detection of nucleic acid targets, has become one of the most important tools in parasite diagnostics. Real-time PCR has simplified and accelerated PCR procedures and has reduced complications associated with traditional PCR, such as cross-contamination. Molecular biology tools, such as PCR, are increasingly relevant to veterinary parasitology. This chapter focuses on the application of real-time PCR for parasite detection and differentiation, exemplified in protozoa, helminthes and arthropods, significant parasites in veterinary medicine and public health.

Keywords: Veterinary parasitology; genotyping; *Toxoplasma gondii*; *Cryptosporidium parvum*; *Theileria equi*; *Strongyloides stercoralis*; *Filariasis*; *Hematozoans*; parasite-host interaction; SYBR® Green; FRET; melting curve analysis.

1. INTRODUCTION

With the development of molecular biology, the approaches to the study of parasites and parasitic diseases have been greatly revolutionized. PCR has become a routine technique employed in clinical diagnostics laboratories. This technology has also been extensively utilized in parasite systematics and epidemiology, parasite-host interactions, vaccine development, as well as parasite functional genomics research. Real-time PCR utilizes a fluorogenic detection system to monitor a continuous measurement of amplified products throughout the reaction. It was first introduced by Higuchi *et al.* to analyze the kinetics of PCR by constructing a system to detect PCR products during the process of their amplification [1, 2]. In this real-time system, the fluorescent dye, ethidium bromide, intercalates preferentially into double-stranded amplification products, which also increases the strength of its fluorescence. The progress of PCR amplification can be continuously monitored in real time by acquiring fluorescence signals in each amplification reaction cycle during segments in which double-stranded DNA is present. Thus, quantitative information about the PCR process is obtained by plotting the intensity of the fluorescence signal versus the cycle number. This is in contrast to conventional PCR methodology in which only qualitative information is obtained at the end point of amplification through the use of gel electrophoresis to visually detect the presence or absence of a specific double stranded DNA product.

The shortcomings of conventional PCR, such as cross-contamination, false-positive signals, or false - negative signals caused by enzyme-inhibitors in the samples limit its clinical application [3]. In contrast,

---

*Address correspondence to Hongzhuan Wu: Department of Biological Sciences, Alabama State University, Montgomery, AL, 36101, USA; Tel: 001-334-229-4498; E-mail: hwu@alasu.edu

real-time PCR detects product directly, without the need to open the reaction tube after PCR completion, thus minimizing post-amplification processing and the risk of sample contamination by products of previous amplifications (product carryover). Real-time PCR has been rapidly adopted for use in many areas of parasitology, such as polymorphism genotyping, detecting and quantifying nucleic acids of parasites, and monitoring the transcription of genes following parasitic infections.

Enzymatic amplification, using species or genus specific sequences, has long been utilized for parasite identification [4, 5]. Methodology is now available to develop PCR tests that simultaneously identify more than one parasite group; several specific primer sets are combined into a single PCR assay, *i.e.*, multiplex PCR [6]. Such a test can substantially simplify analyses of mixed parasite populations. In addition, sensitive, fluorescence-based "real-time" PCR techniques that combine both PCR and fragment analysis are being used to identify parasites and parasitic diseases, as well as to quantitate parasite levels in biological samples. Heightened interest in these methodologies has prompted a review of their application and potential for future use in parasite research.

Parasite differentiation has been challenging with the use of traditional approaches. The advent of molecular techniques, such as restriction fragment length polymorphism (RFLP) of total genomic DNA, has revolutionized parasite differentiation [3]. Repetitive DNA fragments generated from restriction enzyme-digested DNA and separated by agarose gel electrophoresis have been visualized by direct examination of ethidium bromide stained gels. Later, Southern blotting was used to enhance sensitivity using highly abundant radio-labeled probes such as ribosomal RNA (rRNA), cloned ribosomal DNA (rDNA) fragments, or undefined repetitive DNA fragments. However, the use of repetitive DNA probes for parasite identification and differentiation has been gradually replaced by PCR. PCR-based technologies such as PCR-RFLP and random amplified polymorphic DNA (RAPD) have been used extensively for parasite identification and differentiation. The advantages of single-sequence conformational polymorphism (SSCP) [7, 8], multiplex PCR [9, 10], and real-time fluorescence-based PCR [11, 12] have resulted in these techniques being applied more frequently in veterinary parasitology.

2. VETERINARY PARASITE GENOMICS

The traditional approaches to parasitology have served the veterinary field in the diagnosis, vaccine development and screening of antiparasitic drugs, although the outcomes have often been less than desirable. Over the last 50 years, there have been no significant breakthroughs in regard to discovering new classes of broad spectrum antiparasitic drugs. Meanwhile, the variety of drug resistant parasitic infections is increasing, threatening livestock populations in some parts of the world [13, 14]. The knowledge and application of molecular biology are increasingly important in the field of veterinary parasitology, making possible the development of new diagnostic tools, the discovery of antiparasitic drugs, the understanding of antiparasitic resistance, and the development of new vaccines.

The sequencing of the whole and partial parasitic genomes represents a tremendous resource for the research and diagnosis of parasites and their associated diseases. The genome of *Caenorhabditis* (*C.*) *elegans* was the first completed genomic sequence for any multicellular organism [14]. Degenerate primers were designed based on *C. elegans* gene sequences to explore the related genes in parasitic helminthes [15, 16]. Many other ongoing and completed genome sequencing projects have been undertaken by groups such as the Sanger institute (*Ascaris suum* ESTs, *Haemonchus contortus* ESTs, *Trichinella spiralis*, *Plasmodium falciparum*, *Schistosoma mansoni*, *Leishmania* spp., African *trypanosome Trypanosoma brucei*, *Entamoeba histolytica.*, *Toxoplasma gondii*), The Institute for Genomic Research (*Theleria parva*), Washington University (*Neospora parva*) and the University of Minnesota (*Cryptosporidium parvum*) [17-24]. The information generated from genomic sequencing has provided the basis for the development of novel vaccines, the discovery of antiparasitic drugs, and the establishment of new diagnostic approaches.

3. APPLICATION OF QPCR IN VETERINARY PARASITOLOGY

In molecular parasitology, DNA samples for PCR analysis are extracted from different sources of interest (blood, tissues, feces, vectors and environmental samples). The quality and types of the samples significantly influence the efficiency and sensitivity of the DNA extraction, and of the following PCR. The preservation, transportation and processing of the samples and the format for the detection of the PCR target are critical issues for ensuring high-throughput PCR systems. There is also the need to establish a standardized approach to accurately quantify the infectious loads of the parasite. Infection by multiple parasites in a single host is not uncommon, and a reliable PCR with quantitative standards is very useful in interpreting whether the host is suffering from acute or chronic infection, or in a carrier state, and to elucidate the main causative agent for the clinical disease.

For both helminthes and protozoa, the rDNA genes have served as often used targets for the design of primers and probes for PCR-based diagnostics. This is due to the high degree of conservation of these genes and the availability of sequence data for the genes. Another advantage to using the rDNA genes for PCR is that the genes are expressed in multiple copies, thereby increasing the sensitivity of PCR. These rDNA-based PCR systems have been established to detect many parasites of veterinary significance such as *Babesia* species [25-27], *Cryptosporidium* [28], *Hepatozoon* spp. [29] and *Theileria* [30, 31]. Some single copy genes have also been used as the target for PCR, such as the COWP gene of cryptosporidium [32] and the major merozoite antigen of *Theileria* [31] with reasonable sensitivities.

The real-time PCR technique has been used to detect, differentiate, and diagnose parasites of human beings and animals for the past 25 years. Specific FRET real-time PCRs were developed to diagnose *Toxoplasma gondii* [33], *Trypanosoma cruzi* [34], *Cryptosporidium parvum* [35], canine *hematozoan* infections [27, 29], and *Eimeria acervulina* in feces of broiler chickens [36]. A quantitative PCR was developed to detect *Leishmania infantum* in captive wolves and canids [37], and a SYBR® green real-time PCR assay was established for the detection of the nematode *Angiostrongylus vasorum* in definitive and intermediate hosts [38]. Real-time PCR was also developed and evaluated in the quantitative detection of *Babesia caballi* and *Theileria equi* infections in horses from South Africa [39]. A combination of cell culture and quantitative PCR (cc-qPCR) can be used to assess a disinfectant's efficacy on *Cryptosporidium* oocysts under standardized conditions [40].

3.1. *Toxoplasma gondii*

Toxoplasmosis is a worldwide infectious disease caused by the protozoan *Toxoplasma* (*T.*) *gondii*. *T. gondii* is an obligate intracellular coccidian parasite that infects virtually all warm-blooded individuals. The parasite exists in three stages: oocytes in the feces and the tachyzoite and bradyzoite stages in the tissues. Tachyzoites, as the actively multiplying stage, are responsible for the acute infection, and their differentiation into bradyzoites (slowly multiplying stage) correlates with the onset of protective immunity. Bradyzoites are located within cysts, which are believed to persist for life. These quiescent stages are able to reconvert into active tachyzoites when the immune system fails. This reactivation is thought to be the main source of cerebral or disseminated toxoplasmosis in immunocompromised individuals.

The detection of parasites, or detection of evidence of parasites in tissues, is often difficult, which may significantly delay diagnosis and result in a poor prognosis as a consequence. A real-time PCR test, using fluorescence resonance energy transfer hybridization probes was established to detect and quantify *T. gondii* DNA from blood [11]. This PCR test gave reproducible quantitative results over a dynamic range of from 0.75 to 0.75×10^6 parasites per PCR mixture. Highly sensitive quantification of *T. gondii* DNA has also been achieved from pig and mouse tissues [41], from clinical whole blood and amniotic fluid [42, 43], and from cerebrospinal fluid [43]. A LightCycler based FRET PCR was developed to detect *T. gondii* DNA in serum with a threshold of <1 parasite, and to enable the monitoring of *Toxoplasma* reactivation and progression associated with pyrimethamine - clindamycin treatment [11]. The real-time quantitative LightCycler-PCR assay can be used not only to detect the presence of *T. gondii* DNA but also to provide precise evaluations of the parasite load in immunocompromised patients. This PCR test is useful for the monitoring of treatment efficacy and should help provide an understanding of the pathogenesis of *Toxoplasma* reactivation [11].

3.2. *Cryptosporidium parvum*

Cryptosporidium (*C.*) *parvum* is a coccidian protozoan that inhabits the epithelium of the respiratory and digestive tracts of reptiles, birds and mammals. Most species are host specific and *Cryptosporidium* found in reptiles and birds do not infect mammals. *C. parvum* could cause self-limited diarrhea in immunocompetent individuals, and this organism is of particular concern in immunocompromised hosts. In immuncompromised and malnourished individuals, the disease can be severe, prolonged, and life threatening. Although several immunological and molecular methods for detection of *C. parvum* oocysts in stool and environmental samples have been developed, immunomagnetic capture methods have found widespread application, particularly for water monitoring. The detection limits achieved with these systems are typically less than 10 oocysts, although recoveries are affected by the complexity of the matrix from which the oocysts are extracted.

The present laboratory methods for the diagnosis of *C. parvum* by microscopy are generally adequate for samples with high concentrations of oocysts, but are insufficient for the detection of cases of cryptosporidiosis in which only small numbers of oocysts are excreted. Molecular methods for the detection of oocysts in complex mixtures or genotyping of purified oocysts are needed for clinical and epidemiological applications and for water monitoring. Since no specific chemotherapy is available for this organism, early detection of *C. parvum* infections, particularly in immunosuppressed hosts may be critical to providing supportive treatment. Furthermore, the detection of asymptomatic individuals and animals excreting oocysts may be helpful for preventing secondary infections, studying transmission routes, and identifying reservoir hosts. PCR has also been integrated into various genotyping procedures, such as restriction fragment length polymorphism analysis, random amplification methods, and methods detecting conformational polymorphisms. These approaches have been instrumental in advancing our understanding of the taxonomy of the genus *Cryptosporidium* and for studying the transmission of *Cryptosporidium* species and genotypes between various host species.

Several real-time PCR procedures for the detection and genotyping of oocysts of *C. parvum* have been evaluated [35]. A 40-cycle PCR amplification of a 157-bp fragment from the *C. parvum* tubulin gene detected an individual oocyst, which was introduced into the reaction mixture by micromanipulation. SYBR® Green melting curve analysis was used to confirm the specificity of the method when DNA extracted from fecal samples spiked with oocysts was analyzed. *C. parvum* isolates infecting hosts comprise two distinct genotypes, designated type 1 and type 2, and real-time PCR methods with the application of melting curve analysis were developed for discriminating *C. parvum* genotypes. The first method used the same tubulin amplification primers and two fluorescently labeled antisense oligonucleotide probes spanning a 49-bp polymorphic sequence diagnostic for *C. parvum* type 1 and type 2. The second genotyping method used SYBR® Green I fluorescence and targeted a polymorphic coding region within the GP900/poly (T) gene. Both methods discriminated between type 1 and type 2 *C. parvum* on the basis of melting curve analysis. Nested PCR has been established to detect *C. parvum* from human fecal samples and calf fecal samples, and for diagnosis and molecular typing of isolates causing canine cryptosporidiosis [44]. The high sensitivity of real-time PCR will facilitate the detection of asymptomatic carriers. Such information could be valuable for epidemiological studies of human and animal hosts.

3.3. *Theileria equi*

Equine piroplasmosis, caused by *Theileria equi*, has emerged as an important protozoal infection from both veterinary and economic viewpoints. This disease is characterized by fever, anemia, icterus, and hepato- and splenomegaly and mainly occurs in tropical and subtropical areas of the world. Horses that recover from the initial infection often carry the parasites for the rest of their lives. In such cases, it becomes very difficult to detect the parasites in microscopic examination, and the horses become potential disseminators of the parasites. Various ticks, including *Boophilus*, *Hyalomma*, *Dermacentor*, and *Rhipicephalus*, are known as transmission vectors for *T. equi*.

Current diagnosis of equine piroplasmosis relies on microscopic examination and serological assays. Microscopic detection from blood smears has been used as the most standard diagnosis of equine piroplasmosis. However, this technique is relatively laborious when large numbers of blood smear samples need to be simultaneously quantified. Furthermore, it is difficult to detect the parasites from blood smears when the horses are in a state of chronic parasitemia. Several serological assays, such as the indirect fluorescent antibody test (IFAT), ELISA, and the immunochromatographic test (ICT), have been developed for the detection of *T. equi*-specific antibodies. These serologic assays are also restricted due to their antibody-detection limitation and/or potential cross-reactivity to other pathogens. PCR assay has been described as a molecular tool for the genomic detection of *Theileria* parasites. The sensitivity of the PCR assay is higher than that of the classical microscopic examination. A TaqMan real-time PCR assay was established to detect *T. equi* in equine blood samples. The use of this methodology will facilitate the quantitative diagnosis of *T. equi* in clinical laboratories.

3.4. *Strongyloides stercoralis*

Laboratory diagnosis of strongyloidiasis is primarily based on the detection of *Strongyloides stercoralis* larvae by microscopic examination of stool samples. The number of larvae present is very small, especially in chronic infections. Even using formalin-ether concentration, the Baermann method, or coproculture, the detection rate is low and multiple samples have to be examined to achieve adequate sensitivity. Diagnostic techniques play a pivotal role in providing the scaffolding that medical personnel and disease control managers rely on when deciding which infections are the most threatening. Due to a lack of sensitive detection in low-intensity infections, the spatial distribution and burden of many helminthiases are not well understood. This issue is an important reason why helminthiases are often neglected. Increasingly well controlled diseases such as schistosomiasis and lymphatic filariasis, have led to a situation in which the waning morbidity and the dependence on commonly employed, but relatively insensitive, diagnostic techniques conspire to mask the true picture of helminthic diseases. For the control of the neglected group of tropical diseases such as the food-borne trematodiases, schistosomiasis and the common soil-transmitted helminthiases (*i.e.*, ascariasis, hookworm disease, and trichuriasis), there is a tendency to emphasize research and development of new drugs and vaccines, while the critical importance of quality-assured diagnostic tests is receiving far less attention. Moreover, assessment of efficacy of intervention and verification of disease elimination and early warning systems strongly depends on reliable diagnostic assays.

A real-time PCR method targeting the small subunit of the rRNA gene was developed for the detection of *Strongyloides stercoralis* DNA in fecal samples, including an internal control to detect inhibition of the amplification process [45]. The assay was performed on a range of well-defined control samples ($n = 145$), known positive fecal samples ($n = 38$) and fecal samples from a region in northern Ghana where *S. stercoralis* infections are highly endemic ($n = 212$), and achieved 100% specificity and high sensitivity. The use of this assay could facilitate monitoring the prevalence and intensity of *S. stercoralis* infections during helminth intervention programs. Moreover, the use of this assay in diagnostic laboratories could make the introduction of molecular diagnostics feasible in the routine diagnosis of *S. stercoralis* infections, with a two-fold increase in the detection rate as compared with the commonly used Baermann sedimentation method.

3.5. Filariasis

Lymphatic filariasis, a mosquito-borne disease, is a major public health problem, particularly in the tropics and subtropics. It is caused by the nematodes *Wuchereria* (*W.*) *bancrofti*, *Brugia* (*B.*) *malayi*, and *Brugia timori*. Symptoms include acute fever and chill and chronic lymphoedema and hydroceles. Brugian filariasis caused by *B. malayi* and *B. timori* affects 13 million people, mainly in India and Southeast Asia. A principal goal of the Global Program to Eliminate Lymphatic Filariasis (GPELF) is interruption of the transmission of infection. Hence, the availability of efficient and effective diagnostic tools to monitor the presence or absence of filarial larvae in the mosquito vector is particularly important. Entomologic methods for the detection of filarial larvae in mosquito vectors are based on the dissection of mosquitoes. However, these methods are laborious, tedious, and time consuming and carry a low sensitivity and a need for specially trained microscopists.

Many conventional PCR assays have been developed to detect filarial DNA in blood and mosquito vectors. All of these procedures require agarose gel electrophoresis for the analysis. However, determination by gel electrophoresis is slow, has a limited throughput, and is prone to carry-over contamination and subjective results. Recently, effective real-time PCR has greatly improved molecular detection and differential diagnosis of microorganisms belonging to the same genus and has increasingly replaced conventional PCR.

Effective real-time PCR is not only accurate, sensitive, fast, and can quantify specific DNA in biologic samples, but it also differentiates species or strains of several medically pathogenic microorganisms by melting curve analysis. Moreover, this method provides a high-throughput means because it does not need agarose gel electrophoresis for analysis of the amplicons and has a broad dynamic range. The method has great potential for epidemiologic studies and for monitoring elimination programs of infectious agents.

Recently, *W. bancrofti* and *Brugia* spp. DNA have been identified in infected blood and in infected mosquito vectors by a Taqman probe and an Eclipse minor groove binding probe based on real-time PCR. A real-time fluorescence resonance energy transfer (FRET) polymerase chain reaction (PCR) combined with melting curve analysis was developed to detect *B. malayi* DNA in blood-fed mosquitoes [46]. This real-time FRET PCR was based on a fluorescence melting curve analysis of a hybrid formed between amplicons generated from a family of repeated DNA element [47]. The *B. malayi*-infected mosquitoes were differentiated from *W. bancrofti*-infected and uninfected mosquitoes and from genomic DNA of *Dirofilaria immitis*- and *Plasmodium falciparum*-infected blood and leukocytes by their melting temperature. Melting curve analysis produces a rapid, accurate, and sensitive alternative for specific detection of *B. malayi* in mosquitoes, allows high throughput, and can be performed on small samples. This method has the potential for endemic area mapping or for monitoring the effect of brugian filariasis mass treatment programs.

3.6. Hematozoans

Hematozoans such as *Hepatozoon* live in several cell types (such as muscular, hepatic or hematic) of canids and felids, causing signs of acute disease. Infection with *H. canis* is present almost worldwide and can lead to death in dogs. The only exception for this nearly cosmopolitan distribution is North America, where *H. americanum* is the dominant species. In addition, a new *Hepatozoon* spp. has been found in cats from Southern Europe. Its arthropod vector, host range and biogeography remain to be established. A SYBR® Green quantitative PCR assay was established for *Hepatozoon* spp. diagnosis. The molecular approach, consisting of the amplification of a 235 bp fragment of the 18S rRNA gene, is able to detect at least 0.1 fg of parasite DNA. Reproducible quantitative results were obtained over a range of 0.1 ng -0.1 fg of *Hepatozoon* spp. DNA.

Auburn University Molecular Diagnostics Laboratory developed a FRET quantitative PCR to diagnose and differentiate canine hepatozoan [29]. This diagnostic method for detection of *Hepatozoon* spp. DNA integrates nucleic acid extraction with extensive agitation to maximize DNA extraction efficiency. This PCR method amplifies a fragment of the *Hepatozoon* 18S rDNA gene, detects as few as 7 genomic copies of *Hepatozoon* spp. per ml of blood, and simultaneously differentiates between *H. americanum* and *H. canis* amplicons. Interestingly, a surprising 300-fold increase of *H. americanum* 18S rDNA targets occurred after 3 days storage of positive blood specimens at room temperature. The whole blood samples of dogs mostly from the southeastern Unites States identified *H. americanum* in 167 samples (27.2%), *H. canis* in 14 samples (2.3%), and both *H. americanum* and *H. canis* in 14 samples (2.3%) [29].

In addition to *Hepatozoon*, there are other hematozoans transmitted by ticks, such as *Babesia* spp. These piroplasmid protozoa are important pathogens of canids as well. Babesiosis has been traditionally diagnosed by microscopic identification of intra-erythrocytic trophozoites in blood smear, and by serological testing. Due to the fact that the organisms vary or are infrequent in blood smears, molecular detection is the most sensitive and specific method for *Babesia* diagnosis. A quantitative FRET-PCR combined with melting curve analysis was established to amplify a fragment of the *Babesia* spp. 18S rRNA gene, and further differentiates *B. gibsoni*, *B. canis canis/B. canis vogeli*, and *B. canis rossi* [27].

4. FUTURE PERSPECTIVES

Knowledge of molecular biology is increasingly important to the field of veterinary parasitology, such as in the understanding of the parasite-host interactions and the development of novel diagnostic approaches with high sensitivity and specificity. There has been a dramatic evolution, and revolution, in the study of parasites and parasitic diseases with the involvement of molecular approaches. Besides polymerase chain reaction, other molecular advances such as the analysis of EST libraries, the generation of single nucleotide polymorphisms (SNPs), laser capture microdissection and the construction of PCR-based cDNA libraries are being combined with techniques such as microarray and microchip [3]. The field of veterinary parasitology should continue to welcome, embrace and encourage the incorporation of molecular knowledge in our research, clinical work, and veterinary education.

REFERENCES

[1] Higuchi R, Dollinger G, Walsh PS, GriYth R. Simultaneous amplification and detection of specific DNA sequences. Biotechnology 1992; 10: 413-7.

[2] Higuchi R, Fockler C, Dollinger G, Watson R. Kinetic PCR analysis: Real-time monitoring of DNA amplification reactions. Biotechnology 1993; 11: 1026-30.

[3] Zarlenga DS, Higgins J. PCR as a diagnostic and quantitative technique in veterinary parasitology. Vet Parasitol 2001; 101: 215-30.

[4] McKeand JB. Molecular diagnosis of parasitic nematodes. Parasitol 1998; 117: S87-96.

[5] Gasser RB. PCR-based technology in veterinary parasitology. Vet Parasitol 1999; 84, 229-58.

[6] Kaltenboeck B, Wang C. Advances in real-time PCR: application to clinical laboratory diagnostics. Advances in clinical chemistry 2005; 40: 220-49.

[7] Gasser RB, Monti JR. Identification of parasitic nematodes by PCR-SSCP of ITS-2 rDNA. Mol. Cell Probes 1997; 11: 201-9.

[8] Gasser RB, Zhu XQ, Monti JR, Dou L, Cai X, Pozio E PCR-SSCP of rDNA for the identification of *Trichinella* isolates from mainland China. Mol Cell Probes 1998; 12: 27-34.

[9] Zarlenga DS, Chute MB, Martin A, Kapel CMO. A multiplex PCR for unequivocal differentiation of six encapsulated and three non-encapsulated genotypes of *Trichinella*. Int J Parasitol 1999; 29: 141-9.

[10] Zarlenga DS, Chute MB, Gasbarre LC, Boyd PC. A multiplex PCR assay for differentiating economically important gastrointestinal nematodes of cattle. Vet Parasitol 2001; 97: 199-209.

[11] Costa JM, Pautas C, Ernault P, Foulet F, Cordonnier C, Bretagne S. Real-time PCR for diagnosis and follow-up of *Toxoplasma* reactivation after allogeneic stem cell transplantation using fluorescence resonance energy transfer hybridization probes. J Clin Microbiol 2000: 38: 2929-32.

[12] Jauregui LH, Higgins J, Zarlenga DS, Dubey JP, Lunney JK. Development of a real-time PCR assay for the detection of *Toxoplasma gondii* in pig and mouse tissues. J Clin Microbiol 2001; 39: 2065-71.

[13] Zajac AM, Sangster NC, Geary TG. Why veterinarians should care more about parasitology. Parasitol Today 2000; 16(12): 504-6.

[14] Prichard R, Tait A. The role of molecular biology in veterinary parasitology. Vet Parasitol 2001; 98(1-3): 169-94.

[15] Blackhall WJ. Genetic variation and multiple mechanisms of anthelmintic resistance in *Haemonchus contortus*. Ph.D. Thesis. 2000. McGill University, Montreal.

[16] Forrester SG, Hamdan FF, Prichard RK, Beech RN. Cloning, sequencing and developmental expression levels of a novel glutamate-gated chloride channel homologue in the parasitic nematode *Haemonchus contortus*. Biochem Biophys Res Commun 1999; 254: 529-34.

[17] Annon. 2000 www.genome.cvm.umn.edu.

[18] Gardner MJ, Hall N, Fung E, et al. Genome sequence of the human malaria parasite Plasmodium falciparum. Nature 2002; 419 (6906): 498-511.

[19] Berriman M, Ghedin E, Hertz-Fowler C, et al. The genome of the African trypanosome Trypanosoma brucei. Science (New York, N.Y.) 2005; 309 (5733): 416-22.

[20] Ivens AC, Peacock CS, Worthey EA, et al. The genome of the kinetoplastid parasite, *Leishmania* major. Science (New York, N.Y.) 2005; 309 (5733): 436-42.

[21] Loftus B, Anderson I, Davies R, et al. The genome of the protist parasite *Entamoeba histolytica*. Nature 2005; 433 (7028): 865-8.

[22] Haas BJ, Berriman M, Hirai H, Cerqueira GG, Loverde PT, El-Sayed NM. *Schistosoma mansoni* genome: closing in on a final gene set. Experimental parasitol 2007; 117: 225-8

[23] Jeffares DC, Pain A, Berry A, et al. Genome variation and evolution of the malaria parasite Plasmodium falciparum. Nature Genetics 2007; 39: 120-5

[24] Mourier T, Carret C, Kyes S, et al. Genome-wide discovery and verification of novel structured RNAs in *Plasmodium falciparum*. Genome research 2008; 18: 281-92

[25]   Calder JAM, Reddy GR, Chieves L, *et al*. Monitoring *Babesia bovis* infections in cattle by using PCR-based tests. J Clin Microbiol 1996: 34: 2748-55.

[26]   Bashiruddin JB, Camma C, Rebêlo E. Molecular detection of *Babesia equi* and *Babesia caballi* in horse blood by PCR amplification of part of the 16S rRNA gene. Vet Parasitol 1999; 84: 75-83.

[27]   Wang C, Ahluwalia SK, Li Y, *et al*. Frequency and therapy monitoring of canine *babesia spp*. Infection by high-resolution melting curve quantitative FRET-PCR. Vet Parasitol 2010; 168: 11-8.

[28]   Morgan UM, Thompson RC. Molecular detection of parasitic protozoa. Parasitol 1998; 117 (Suppl.): S73-85.

[29]   Li Y, Wang C, Allen KE, *et al*. Diagnosis of canine *Hepatozoon spp*. infection by quantitative PCR. Vet Parasitol 2008; 157: 50-8.

[30]   Bishop R, Allsopp B, Spooner P, Sohanpal B, Morzaria S, Gobright E. *Theileria*: improved species discrimination using oligonucleotides derived from large-subunit ribosomal RNA sequences. Exp Parasitol 1995; 80: 107-15.

[31]   d'Oliveira C, van derWeide M, Habela MA, Jacquiet P, Jongejan F. Detection of *Theileria annulata* in blood samples of carrier cattle by PCR. J Clin Microbiol 1995; 33: 2665-9.

[32]   Spano F, Putignani L, McLauchlin J, Casemore DP, Crisanti A. PCR-RFLP analysis of the *Cryptosporidium* oocysts wall protein (COWP) gene discriminates between *C. wrairi* and *C. parvum* isolates of human and animal origin. FEMS Microbiol Lett 1997; 150: 209-17.

[33]   Costa JM, Pautas C, Ernault P, Foulet F, Cordonnier C, Bretagne S. Real-time PCR for diagnosis and follow-up of *Toxoplasma* reactivation after allogeneic stem cell transplantation using fluorescence resonance energy transfer hybridization probes. J Clin Microbiol 2000; 38: 2929-32.

[34]   Britto C, Cardoso A, Silveira C, Macedo V, Fernandes O. Polymerase chain reaction (PCR) as a laboratory tool for the evaluation of the parasitological cure in Chagas disease after specific treatment. Medicina Buenos Aires 1999; 59: 176-8.

[35]   Tanriverdi S, Tanyeli A, Baslamisli F, *et al*. Detection and genotyping of oocysts of *Cryptosporidium parvum* by real-time PCR and melting curve analysis. J Clin Microbiol 2002; 40: 3237-44.

[36]   Velkers FC, Blake DP, Graat EA, *et al*. Quantification of *Eimeria aceria acervulina* in faeces of broilers: Comparison of McMaster oocyst counts from 24h faecal collections and single droppings to real-time PCR from cloacal swabs. Vet parasitol 2010; 169: 1-7.

[37]   Aoun O, Mary C, Roqueplo CD , Marie´ J,Terrier O , Levieuge A , Davoust BCanine leishmaniasis in south-east of France: Screening of *Leishmania infantum* antibodies (western blotting, ELISA) and parasitaemia levels by PCR quantification. Vet Parasitol 2009; 166: 27-31.

[38]   Jefferies R, Morgan ER, Shaw SE. A SYBR green real-time PCR assay for the detection of the nematode *Angiostrongylus vasorum* in definitive and intermediate hosts. Vet Parasitol 2009; 166: 112-8.

[39]   Bhoora R, Quan M, Fransen L, *et al*. Development and evaluation of real-time PCR assays for the quantitative detection of *Babesia caballi* and *Theileria equi* infections in horses from South Africa. Vet Parasitol 2009; 168: 201-11.

[40]   Shahiduzzaman M, Dyachenko V, Keidel J, Schmaschke R, Daugshies A. Combination of cell culture and quantitative PCR (cc-qPCR) to assess disinfectants efficacy on *Cryptosporidium oocysts* under standardize conditions. Vet Parasitol 2010; 167: 43-9.

[41]   Jauregui, L.H. *et al*. (2001) Development of a real-time PCR assay for detection of *Toxoplasma gondii* in pig and mouse tissues. J.Clin.Microbiol.39, 2065-2071.

[42]   Lin, M.H. Higgins J, Zarlenga D, Dubey JP, Lunney JK. Real-time PCR for quantitative detection of *Toxoplasma gondii*. J Clin Microbiol.2000; 38: 4121-5.

[43]   Kupferschmidt, O. Krüger D, Held TK, Ellerbrok H, Siegert W, Janitschke K. Quantitative detection of *Toxoplasma gondii* DNA in human body fluids by TaqMan polymerase chain reaction. Clin Microbiol.Infect.2001; 7: 120-4.

[44]   Scorza AV, Brewer MM, Lappin MR. Polymerase chain reaction for the detection of *Cryptosporidium* spp. in cat feces. J. Parasitol. 2003; 89(2): 423-6.

[45]   Verweija JJ, Canalesb M, Polmanc K, *et al*. Molecular diagnosis of *Strongyloides stercoralis*in faecal samples using real-time PCR. Trans R Soc Trop Med Hyg 2009; 103: 342-6.

[46]   Thanchomnang T, Intapan PM, Lulitanond V, *et al*. Rapid detection of *Brugia malayi* in mosquito vectors using a real-time fluorescence resonance energy transfer PCR and melting curve analysis. Am J Trop Med Hyg 2008; 78: 509-13.

[47]   Criado-Fornelio A, Buling A, Cunha-Filho NA, *et al*. Rapid detection of *Brugia malayi* in mosquito vectors using a real-time fluorescence resonance energy transfer PCR and melting curve analysis. Vet Parasitol 2007; 150: 352-6.

CHAPTER 7

PCR and Veterinary Cancer Diagnostics

Fabio Gentilini*, Maria Elena Turba and Claudia Calzolari

Department of Veterinary Medical Sciences, University of Bologna, Italy

Abstract: Recent advances in molecular biology are providing new opportunities for the diagnosis, prognosis and treatment of cancer. At two decades from its discovery, PCR with its hundreds of variants and improvements is still the keystone of molecular techniques which gave rise to such a revolution. Novel methods and techniques have been introduced in recent years which promise further breakthroughs. Unfortunately, in veterinary medicine, the unaffordable cost of new instrumentation has delayed their application and practical use in clinical settings. Nevertheless, thanks to PCR, the exciting era of molecular medicine has also begun in veterinary medicine. Due to the limited space available, this chapter cannot deal with all the potential applications of PCR in the cancer battlefield. Thus, the authors have focused their attention on those PCR applications concerned with for the diagnosis and prognosis of cancer in pets which are already currently available, albeit not diffusely, at both academic and private laboratories around the world. In some cases, such as in *c-KIT* somatic mutations, for the first time in veterinary medicine, a consensus panel of specialists has recommended the inclusion of a molecular assay in the staging work-up of a neoplastic disease (canine mast cell tumors). In addition to the role of the molecular biologist in developing, implementing and refining the molecular classification for routine clinical practice, it is necessary to discover and validate new targets able to provide accurate information regarding diagnosis, prognosis, treatment resistance, susceptibility or predisposition to toxicity, or the prediction of a therapeutic response.

Keywords: PCR; cancer diagnosis; prognosis; lymphoma; Ig/TCR gene rearrangements; minimal residual disease; mast cell tumors; somatic mutations; telomerase.

In the previous chapters, the reader was provided with almost all the fundamental technical aspects of PCR. In the present chapter, such aspects will be cited in the context of cancer diagnostics, emphasizing the clinical implications and meaning. Technical details will be provided only for complex and very particular PCR applications, such as hairpin-shaped primers or COLD-PCR. The reader is referred to the previous chapters or specific literature for further details.

1. PCR IN THE DIAGNOSIS AND PROGNOSIS OF LYMPHOPROLIFERATIVE DISORDERS IN VETERINARY MEDICINE

1.1. Ig/TCR Gene Rearrangements for the Diagnosis of Canine Lymphoma

1.1.1. Introduction

The diagnosis of canine lymphoid malignancies is achieved using traditional morphologic techniques alone or in conjunction with immunophenotyping assays. In some cases, however, the distinction between malignant and reactive lymphoproliferations can be a diagnostic challenge; these pathologies are found either at the early stage of a lymphoid neoplasia when cancer cells represent only a small subset of cells, in chronic indolent lymphomas or even in a specimen failing to adequately depict the attributes of the lesions because it was harvested using minimally invasive collection techniques such as an endoscopic biopsy. In all these cases, molecular tools may support traditional diagnostic techniques by overcoming many of the uncertainties. Molecular tools are mainly based on the PCR amplification of tumor markers such as recurrent chromosomal aberrations or clonal expansion of immunoglobulins (Igs) and/or T-Cell Receptor (TCR) gene rearrangements. The former is extensively exploited in human medicine and the identification of specific translocations or fusion genes has even been recently integrated in the WHO classification of lymphoid malignancies due to their diagnostic contribution and prognostic features [1].

Address correspondence to Fabio Gentilini: Associate Professor, Department of Veterinary Medical Sciences, University of Bologna, *via* Tolara di Sopra 50, 40064, Ozzano dell'Emilia (Bo), Italy; Tel: +39 051 2097978; Fax +39 051 2097593; E-mail: fabio.gentilini@unibo.it

Chengming Wang, Bernhard Kaltenboeck and Mark D. Freeman (Eds)

Molecular clonality detection is based on the assumption that neoplasia is a clonal expansion of cells derived from a single precursor cell and is therefore characterized by a distinct genetic sequence which could be used as a fingerprint of each malignant lymphoproliferation [2]. Because of their inherent diversity, immunoglobulin and TCR gene loci are useful markers for establishing lymphoid clonal expansion. To better understand how PCR works in detecting clonality and why the Ig and TCR loci are physiologically so different in sequence and length enabling them to be used as a marker of clonality, the following paragraph will focus on the genetic organization of lymphocytes and on the different mechanisms of diversity.

1.1.2. Genetic Organization of Immunoglobulins and TCR Chain Loci and Mechanisms of Diversity

B and T lymphocytes are capable of recognizing a nearly infinite assortment of antigens because of the extreme variability of their Ig and TCR antigen receptor binding sites. These proteins are heterodimers: Igs are composed of two heavy (H) and two light (L) chains (λ and κ) while TCR is composed of two chains ($\alpha\beta$ or $\gamma\delta$ chains). Both Ig and TCR encompass a variable domain responsible for antigen recognition and binding, and one or more constant domain(s) providing the effector response. The V domains include the variable regions of the heavy and light chains (V_H and V_L) characterized by sequence heterogeneity while the constant domains are composed of one or more constant regions (C_H and C_L) which are highly homologous [3]. The V and C regions are coded by V and C genes, respectively, which are located in a unique gene family locus spanning hundreds of kilobases in the germ line DNA. Each gene family is made up of different V and one or more C gene segments. Furthermore, in different Ig chains, additional gene segments, such as D (diversity) and J (joining), may be variably interposed between the V and C gene segments. The V, D, J and C region genes of each locus are not immediately adjacent to each other in the germ line gene organization. Therefore, they do not represent an independent expression unit but need to be "fused" during a recombination process, called gene rearrangement, which occurs during early lymphoid development. During the rearrangement, different V, (D), J and C gene segments are juxtaposed to create a single expression unit which codes for the final polypeptide.

Each gene family is located in a different chromosome locus. In human beings, there are three different loci encoding for immunoglobulins, one for the heavy chain and two for the light chains [4]. In dogs the heavy chain locus is located on chromosome 8 [5]. The TCR genetic organization is composed of four different loci for the four polypeptidic chains (α, β, γ, δ). In dogs, the TCRγ locus on chromosome 5 [6] has recently been identified.

The sequence variability within every V gene segment is not equally distributed; each V segment is composed of three conserved nucleotide sequences (framework regions, FR I, II and III) and by three hypervariable sequences, the complementary determining regions (CDRs I, II, III) which code for the antigen binding sites. In particular, CDRIII is located at the junction site of the V segment (between the V and D segments) and is responsible for encoding the hypervariable region of the binding site [2] (Fig. **1**).

There are three different mechanisms of diversity: somatic recombination (also called V(D)J rearrangement), junctional diversity and somatic hypermutation [4]. The V(D)J rearrangement is due to the recombination process which leads to the assembling of the expression unit for the variable domain of the protein by linking one of the different V, D and J segments; this mechanism occurs during early lymphoid development and is mediated by the recombinase enzyme complex (RAG) which recognizes the gene segments at the recombination signal sequences, cleaves the DNA double strand and finally repairs the double strand break using a non-homologous end-joining mechanism [7]. The many different possible combinations of V(D)J segments represent the so-called combinatorial repertoire. During the V(D)J rearrangements at the junction sites between gene segments, a second mechanism of diversity occurs: the random insertion or deletion of nucleotides resulting in highly diverse junctional sequences which significantly contribute to generating a sequence length heterogeneity (junctional diversity) (Fig. **1**). Finally, in germinal centers, the B-mature lymphocytes undergo an additional mechanism of diversity called somatic hypermutation; during antigen recognition, mediated by T-helper lymphocytes, single-nucleotide mutations and insertions/deletions occur extending the Ig repertoire. Only those random mutations which increase the affinity for the antigen are further expanded; the others result in apoptosis [4, 7]. All these mechanisms of diversity create a unique fingerprint, mainly determined by the extreme variability of CDRIII which could be used as a marker of clonality.

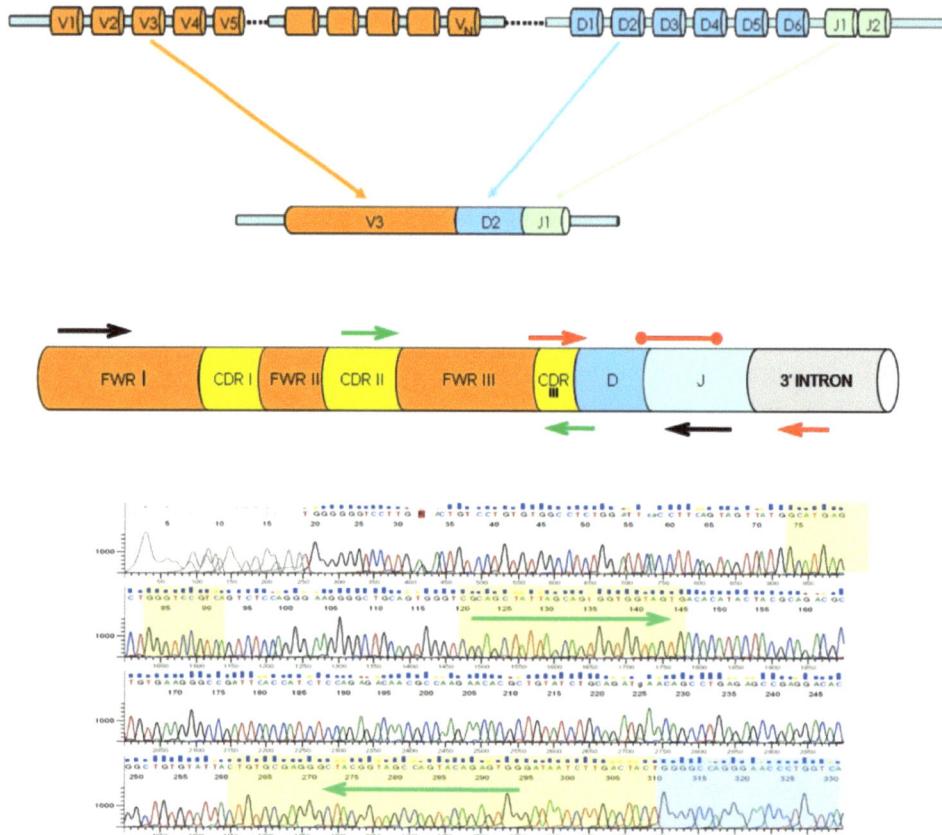

Figure 1: Germline DNA rearranged in lymphoid cells. Primer positions are schematically illustrated. The sequencing of a neoplastic clone amplified with a forward primer annealing in framework region I is used to design clone specific primers.

1.1.3. Clonal Detection Using Antigen Receptor Gene Rearrangements by PCR

Clonality assays in canine and feline lymphoid malignancies rely on V(D)J rearrangement assessment by PCR amplification of the heavy chain locus of the Ig (IgH) and the γ locus of TCR. Assays for pets should take into account the specific differences in immunoglobulin genetics between humans and dogs; for instance, the possibility of using the light chains as a marker of clonality is a human prerogative because λ and κ chains are expressed in equal ratio. In this context, the imbalance due to the predominance of one of the two chains could represent a marker of neoplasia. Unfortunately, in canine immunoglobulins, the lambda light chains are preferentially expressed in a 9 to 1 ratio under normal conditions [8], hampering the inclusion of IgL as a target of clonality assays. Conversely, the γ locus of TCR was preferred because previous studies in human medicine suggested that TCRγ is easier to amplify than the TCRβ locus whereas the TCRα gene structure is too complex and TCRδ is often deleted in mature T cells [2, 9].

Clonality markers may be targeted with different molecular techniques. Southern blotting methods represented the gold standard until almost 2 decades ago when they were replaced by PCR which had several advantages; PCR is rapid, sensitive, less labor-intensive and requires a very small amount of DNA. The first PCR method described in veterinary medicine was initially described by Vernau and Moore in 1999 and was further improved and validated by Burnett and colleagues in 2003, in the era preceding complete canine genome sequencing. The authors used the IgH and TCRγ targets to design a consensus primer after extensive cloning of mRNA and accurate *in silico* alignments of the cloned sequences. Two sets of consensus probes annealing within the putative FRIII region of the V_H segments (forward primer) and within the J_H segment (reverse primer) were selected for amplifying the intercalated IgH CDRIII region. TCR γ rearrangements were amplified with only one set of primers including one degenerate forward primer and 2 different reverse primers [10, 11]. An additional set of primers targeting IgH and TCRγ was subsequently reported [5, 6, 12]. These sets of primers amplify only the rearranged DNA of lymphoid cells and not the germ line DNA of somatic cells because the gene segments are too distant for PCR amplification. The normal or reactive/hyperplastic population of lymphocytes carries a multitude of gene rearrangements, each one with a definite length. The amplification of the DNA purified from such lymphoid cells yields PCR products of different lengths which could be resolved in high resolution polyacrylamide gels with characteristic ladder or smear patterns (polyclonal pattern). On the other hand, neoplastic populations include clonally expanded cells with the same rearrangement either Ig in B cell lymphoma or TCRγ in T cell lymphoma. The neoplastic clone rearrangements are amplified by a consensus primer yielding a unique or prevalent amplicon which appears as a single discrete band after gel electrophoresis (monoclonal pattern). Finally, some populations of lymphocytes may present an intermediate pattern characterized by a few discrete bands (oligoclonal pattern) [11]. The significance of oligoclonal patterns is uncertain. Clearly, oligoclonal pattern definition depends on the visualization techniques and their relevance will be discussed below.

The detection of clonal expansion of Ig or TCRγ is suggestive of B and T cell neoplasia, respectively. Lineage assignment of a lymphoid neoplasia throughout PCR has been proposed as an adjunctive potential application even though a subset of the high grade lymphomas may have a cross-lineage rearrangement [9, 13]. Consequently, the phenotype of lymphoid malignancies and its prognostic implication as assessed using immunological methods may or may not correspond to the PCR finding. Indeed, one of the main advantages of a clonality assay using PCR is the possibility of examining a wide range of biological samples: lymph nodes, bone marrow, spleen, liver, intestine, nasal mucosa, cutaneous lesions and peripheral blood; cerebrospinal fluid, ocular fluids and effusions may also be assessed after sampling with minimally invasive techniques such as centesis and fine needle aspiration. Specimens suitable for clonality assay are freshly collected cells, unstained smears, stained smears or even formalin-fixed paraffin-embedded tissues. Regarding the latter, modified DNA extraction methods are strongly recommended because formalin induces both DNA/protein cross-linking and DNA fragmentation, hampering the purification of high quality DNA with a molecular weight greater than 200 bp [14].

PCR clonality assays have shown their robustness and overall accuracy in many studies of canine lymphomas [5, 11, 12, 16, 17]. Indeed, a PCR clonality assay can detect neoplastic cells in the peripheral blood of the majority of lymphoma-affected dogs with a higher sensitivity than morphological techniques, although this finding does not predict the disease-free interval or survival [15, 17].

Nevertheless, false positive and false negative results may occur and should be carefully taken into account. False negative results might be due to polymorphism or mutations occurring at the consensus sequences; the simultaneous use of multiple primer sets or patient-specific primer designs are currently being used for overcoming this occurrence. [12, 19]. False negative results may also be caused by the absence of adequately rearranged DNA in the sample. This is because either the neoplastic transformation occurred in an early precursor which has not yet undergone rearrangement or in developed lymphoid cells which do not undergo Ig/TCR rearrangement, such as Natural Killer cells. Finally, false negative results may occur in samples with neoplastic cells (scattered among a majority of normal cells) which are below the limit of detection. In clonality assays, the threshold is set at approximately 1-5% of cells [11, 19, 20]. A lower detection limit is hampered by the phenomenon of primer competition on the consensus sequences of both neoplastic cells and normal cells leading to the erroneous appearance of polyclonal patterns also in the presence of neoplasias whenever there is a prevalence of normal cells. It has been postulated that false negative results are also possible due to low quality highly fragmented DNA [12].

False positive results may be caused by hyperplastic/reactive lymphoproliferations leading to reduced diversity of the Ig/TCR repertoire caused by antigen-driven subclone selection as seen in some cases of canine ehrlichiosis or Lyme disease [11, 20]. Caution is also necessary in interpreting PCR results from samples containing very few lymphocytes; in fact, the oligoclonal/monoclonal pattern resulting from the amplification process of only a few lymphoid cells may resemble a clonal population. False positive findings in terms of lymphoid assignment may result from some myeloid neoplasms, such as acute myeloid leukaemia with IgH rearrangements [20]. False positive results are more likely when evaluating canine TCRγ rearrangements due to the limited repertoire of gene segments involved. Finally, false positive results have also been observed secondary to lymph-node necrosis [21] and eosin staining of samples run with capillary electrophoresis (see below; [21]).

1.1.4. Advances in Clonal Rearrangements Detection

The standard techniques of electrophoresis on polyacrylamide gels are not capable of discriminating between DNA sequences having minimal length differences and for that reason heteroduplex analysis is mandatory to confirm positive results [18, 19]. An alternative PCR downstream application for detecting clonality is GeneScanning analysis which is used extensively in human medicine for diagnostic and prognostic purposes [18]. GeneScanning is a capillary electrophoresis-based technique which enables automated, high-throughput fragment analysis. The principle of the detection system is that the nucleotide probes are 5' labelled with fluorescent dyes which are activated by a laser beam. GeneScanning couples highly discriminative features with highly sensitive fluorogenic detection of the amplicons. The labelled PCR products are visualized as colourful peaks with size and heights calculated by specific software and are compared to an internal size standard embedded in the electropherogram (Fig. **2**). Genescanning exhibits many advantages over other techniques; it has 1) an elevated discriminative power enabling the differentiation of PCR products differing in size by only one nucleotide 2) high sensitivity allowing the detection and analysis of a very small amount of PCR products 3) elevated processivity due to its inherent possibility of simultaneously analyzing different PCR products in the same run (multiplex runs). A drawback is represented by the fact that the instrumentation is quite expensive, limiting its availability in clinical practice. GeneScanning analysis exploiting the use of many primer sets previously [5, 11] and newly described achieved a noticeable sensitivity of 97.9% when used to assess clonality in a sample of 96 cases of canine lymphomas [12]. Two new primer sets were designed within the conserved FRI region, at 5' upstream of the FRIII region in which the ones previously reported were designed. Genescanning analysis relies on pattern interpretation rather than on the presence or absence of a signal [12], and the pattern results must be interpreted according to criteria obtained from the human counterpart [19], as seen in Fig. **2**. Fluorescent signals may appear arranged in polyclonal, oligoclonal and monoclonal patterns. A difference exists in olicoclonal pattern interpretation when compared with other visualization techniques; oligoclonal patterns were considered negative since only the monoclonal or biclonal pattern should be confidently interpreted as a positive result. Polyclonal patterns, as documented by GeneScanning, frequently include the typical "Gaussian-like curves" though more complex patterns are not unusual. For instance, monoclonal peaks could be embedded with polyclonal background peaks. In these cases, according to previously reported criteria, the samples should be considered monoclonal only if a peak two-fold higher than any other peak is present [12, 19]. Atypical patterns could arise from samples containing a low percentage of malignant cells in polyclonal reactive background cells, but may also represent DNA degradation artifacts which can even mask an underlying monoclonal population [12]. For this reason, DNA integrity must always be checked, especially in samples stored long-term or formalin-fixed paraffin-embedded tissues in order to avoid false negative results [12, 19].

1.1.5. Clonality Assessment in Feline Lymphoid Malignancies

The feline Ig and TCR γ gene loci have also recently been investigated in order to establish clonality assays. However, the results are not yet completely suitable for clinical purposes. Indeed, multiple primer sets were able to identify clonal lymphoid cells in a maximum of 70% of cases of feline lymphoma/leukaemia [22-24]. These unsatisfactory results differ considerably from their canine and human counterparts and can be explained by a not yet fully known molecular structure of feline Ig/TCR loci. In particular, the Ig region showed many differences in sequences which can be clustered in at least five subgroups. Within these subgroups, IgV3 is the most represented as it is in humans. J segment variability also requires the use of more reverse primers or degenerate ones [22]. Data are accumulating rapidly and suitable molecular assays with improved accuracy will likely be available soon [24, 25].

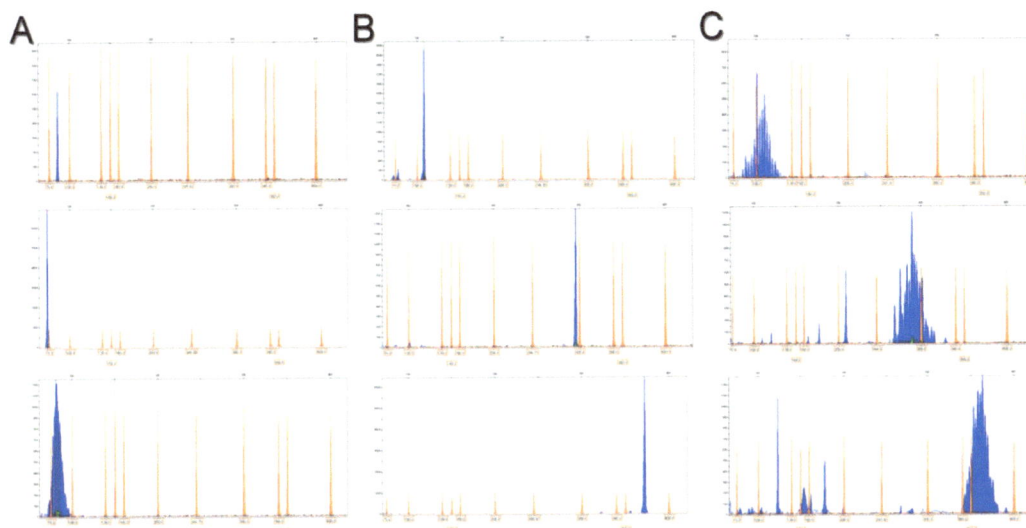

Figure 2: GeneScanning assay: Typical results of TCR (column A) and Ig (columns B and C) clonality assays. A) monoclonal (upper and middle graphs) and polyclonal (lower graphs) peaks of the TCRγ target; B) monoclonal peaks of the Ig target. Upper graph showing forward primer annealing in framework region III of the V gene segment. Middle and lower graphs showing forward primer annealing in framework region I. C) the same target as in B yielding polyclonal patterns.

1.1.6. Future Perspectives

Molecular detection of clonality by PCR noticeably assists the diagnosis of canine lymphoid proliferations. Hopefully, the possibility of detecting neoplastic clones even if they represent a small percentage of the population could represent an adjunctive tool enabling an earlier diagnosis of lymphoid neoplasia.

Besides its role in Ig/TCR gene rearrangement clonality assays, PCR plays a pivotal role in human medicine for its diagnostic and prognostic value which also greatly utilizes the knowledge of highly recurrent chromosomal aberrations. The PCR assessment and monitoring of such genetic alterations could have a strong impact on the molecular identification and quantification of minimal residual disease throughout the course of lymphoid neoplasia. Moreover, the assessment of chromosomal aberration could be used to re-classify lymphoproliferative disorders improving the definition of their biological behaviour and prognosis. Many efforts are ongoing to establish the molecular classification of lymphoid neoplasia in pets and results are expected in forthcoming years.

1.2. Minimal Residual Disease Assessment Using Real-Time Quantitative PCR

Besides its role as an adjuvant diagnostic tool for lymphoproliferative disorders, PCR and, in particular, real-time quantitative PCR (RT-qPCR), is extensively used to monitor minimal residual disease (MRD). Indeed, the cure rates of hematological neoplasias have dramatically increased over time but, unfortunately, many patients still have recurrences and ultimately die. It has become clear that the subset of patients who relapse are those in whom a few neoplastic cells remain after therapy. Since these few cells can only be identified by molecular tools, the concept of molecular remission has been used to identify those patients who have attained a probable cure and who either do not have recurrences or have recurrences much later. Overall, in human medicine, RT-qPCR for MRD assessment is used for risk group stratification and is the best detection choice for early treatment modification, outcome prediction and objective evaluation of experimental new treatments.

Different types of molecular targets can be used in RT-qPCR. Whenever present, specific chromosomal aberrations can be detected by PCR, exploiting the unique sequence target of the fusion genes. Secondly, the elevated sensitivity and specificity of PCR may be used for targeting Ig/TCR gene rearrangements (see above). Currently, in veterinary medicine, knowledge of specific and recurrent chromosome aberrations in hematological neoplasias is rapidly increasing but still scarce [26]; therefore, the possibility of using the Ig/TCR gene rearrangements appears more reliable for veterinary patients.

The first descriptions of PCR in assessing MRD date back to the late 1980's - early 1990's. Sensitivity and specificity were achieved by coupling PCR with time-consuming and labor intensive southern blotting radioactive hybridization probes [27]. Since then, more rapid but equally accurate PCR techniques have been validated and described.

A very simple approach entails the use of a qualitative PCR, using the same consensus primer as that used for diagnosing samples collected in the remission phase. However, consensus primers compete at the same target sequences shared between neoplastic and non-neoplastic clones, markedly reducing the ability of evidencing the clonal lymphocytes. It has been estimated that this assay has a limit of detection reaching approximately 1%. Nevertheless, a study attempting to verify whether the same consensus primer used for diagnosis would be reliable for monitoring MRD by sampling lymph nodes after achieving complete remission, was unrewarding [28]. Indeed, the study showed that even dogs that had a post-therapy negative assay might have early recurrences and, in all cases, the post-therapy negative result is not correlated to a delayed relapse [28]. There are two main drawbacks involved in explaining these findings: first, the assay using the consensus primer is not sensitive enough and second, the difficulties in sampling lymph nodes in remission yielded the possibility of negative results due to the sampling of non-lymphoid cells because the so-called DNA control simply represents a PCR amplification control and does not guarantee that lymphoid DNA is present.

The sensitivity of PCR targeting Ig/TCR gene rearrangements can be dramatically increased by means of clone-specific tailored oligonucleotides. After PCR amplification and the sequencing of the Ig/TCR clonal gene rearrangement, such oligos are designed to align themselves in hypervariable sequences of the neoplastic clone. The PCR may be quantitative and carried out in real-time. Ideally, RT-qPCR targeting clone specific sequences should be able to distinguish even a few neoplastic cells interspersed in a prevailing background of normal or reactive cells, each one carrying a specific rearrangement of an unknown and unpredictable sequence. Furthermore, the assay should be robust, reproducible and have a wide dynamic range. Many strategies exploiting the specificity of allele-specific (AS) oligos have consistently been used in RT-qPCR for MRD assessment in human beings. Both consensus forward and reverse primers coupled with patient-specific internal probes (either hydrolysis or hybridization probes) complementary to junctional sequences, or AS primers coupled with consensus internal probes or mixed approaches have been reported to achieve adequate sensitivity and specificity. With clone-specific PCR, a single neoplastic cell scattered among 1×10^6 normal cells can be detected.

Many of the techniques described in human medicine entail the extensive use of fluorogenic probes. Fluorescent-labelled patient-specific probes confer high sensitivity and specificity but, in turn, are very costly. To reduce the unaffordable cost of patient-specific fluorescent probes, two different strategies have also been described in veterinary medicine and, in particular, in dogs (Fig. **1**). In one study, RT-qPCR amplifies the Ig region using consensus fluorescent-labelled probe and reverse primer together with an AS forward primer targeting the hypervariable CDRIII region [29]. In the other study, an assay using AS forward and reverse primers targeting CDRII and CDRIII, respectively, and the intercalating dye SYBR® Green I has been validated [30]. Notwithstanding this, the assay used two AS primers. To attain absolute specificity, the modification of both primers in the hairpin-shaped configuration would be crucial. Indeed, the use of intercalating dye and of consensus labeled probes in this particular context has raised concerns about the specificity due to the fact that the oligos should anneal in a reaction mix containing thousands of similar, unpredictable and unknown sequences. Hairpin-shaped primers have been described to type single nucleotide polymorphisms and are made by adding a short 5'oligonucleotide tail, reverse and complementary to the 3'primer so that the primer can assume a thermodynamically stable hairpin shape [31]. Extensive validation experiments have shown that the hairpin-modification allows the attainment of adequate specificity and sensitivity in a low cost assay. Furthermore, since the CDR II and CDR III regions are far enough from the consensus primer sites, their complete sequencing, needed for clone-specific primer design, could be attained by simple direct sequencing without cloning. For the rapid and complete translation of the molecular assay in the clinical context, this step is also very important.

One of these assays has already been used in a clinical setting [29] yielding very interesting findings. Indeed, a very low level of MRD was observed in all peripheral blood samples, even those collected after attaining complete remission. This finding differs from analogous studies on human Non-Hodgkin lymphomas (NHLs). In NHLs, unlike leukemia, MRD in the peripheral blood during complete remission, even in those patients who have a recurrence, could be detected only inconsistently. Further studies are needed to confirm these findings in dogs.

Provided that some further validation is carried out, this molecular assay will hopefully be available for canine and also feline veterinary practitioners in coming years. The possibility of monitoring the course of the disease and of eventually modifying treatment well before a clinically evident recurrence would lead to the improved effectiveness of treatments and prolonged survival times. Moreover, the assay would be of pivotal importance in screening for the absence of neoplastic cells; whether they are autologous stem cells or bone marrow, transplantation protocols will be set up for veterinary patients with the aim of curing the patient. Finally, MRD assessment would allow the objective and early evaluation of experimental new drugs or protocols without the need for a long study period to evaluate the disease-free interval or even survival.

2. *c-KIT* SOMATIC MUTATIONS AS A PROGNOSTIC MARKER OF MAST CELL TUMORS IN DOGS AND CATS

2.1. Mast Cell Tumor Biology and Clinical Aspects

Mast cell tumours (MCTs) are among the most common neoplasias found in veterinary medicine and their incidence in both dogs and cats is far higher than in humans. However, the same pathway alterations of feline and canine MCTs (*i.e.,* KIT protein kinase constitutive activation) are also present in aggressive systemic mastocytosis and mast cell leukemia in humans [32-38]. This feature has aroused interest in canine and feline MCTs in translational medicine leading to extensive investigation of the molecular mechanism underlying the tumorigenesis, and at evaluating molecularly targeted drugs. Thus, also thanks to PCR, we now have a better understanding of MCTs than we did a few years ago.

The course of the disease can range from benign to aggressive behaviour with early metastasis to the regional lymph nodes or distant organs. Malignant lesions are more frequent in dogs than in cats. Probably for this reason, many molecular markers were first investigated in the former. We herein primarily refer to canine MCTs. In dogs, the course of the disease is correlated to histological grade according to Patnaik *et al.* [39, 40] as poorly differentiated grade III tumours are highly malignant. Since they are also reliable for prognosis, tumor location and WHO clinical stage should be assessed in the work-up procedures while cellular proliferative markers ("growth rate" markers, such as Ki67, and markers of the rate of cell proliferation, such as argyrophilic nucleolar organizing regions [AgNOR]), and KIT expression patterns should be assessed by the pathologist using immunohistology. Three different patterns of KIT expression have been found: a cytoplasmic diffuse pattern, a membranous pattern with immunostaining located on the cell surface and a cytoplasmic perinuclear pattern. The latter pattern was correlated to the presence of somatic mutations and carries some relevant prognostic information [41, 42].

Based on histological grade and clinical staging, treatment includes surgery alone or various combinations of surgery, radiotherapy and chemotherapy with vinblastine, lomustine and prednisone. Very recently, molecular targeted therapy with tyrosin kinase inhibitors has been attempted in dogs with partial success [38, 43-47]. Nevertheless, in many cases, the biologic behaviour of these tumours is unpredictable and the prognosis is challenging. Thus, a consensus panel of veterinary oncologists has recently recommended the evaluation of c-*KIT* somatic mutations in the diagnostic and prognostic work-up of canine MCTs (Joint Meeting of ESVONC and VCS, Copenhagen 2008).

2.2. Usefulness of the Assessment of *c-KIT* Mutational Status in Canine and Feline MCT

The *c-KIT* is a protooncogene which encodes for KIT protein, a type III tyrosine kinase receptor activated by the binding of its ligand stem cell factor (SCF). KIT is expressed by mast cells and mast cell progenitors, germ cells and the interstitial cells of Cajal in the gastrointestinal tract. Upon ligand binding, the intracellular kinase domain activates downstream signalling pathways pivotal for mast cell growth and differentiation [48]. The receptor is composed of an extracellular domain including 5 immunoglobulin-like folds (spanning from exon 2 to 9), a transmembrane domain (coded by exon 10), an intracellular juxtamembrane domain (exon 11) and two intracellular kinase domains (spanning from exon 11 to exon 20).

Each KIT domain plays a distinct role in biological functions. It has been documented that somatic mutations in canine *c-KIT* at exons 8, 9, 11, 12 and 17 lead to ligand independent kinase activation [33, 34, 43, 45, 48, 49]. The constitutive activation of mutated *c-KIT* transfected cells may be reversed by tyrosine kinase inhibitors (TKI) [45, 49].

Analysis of the currently available literature showed that, overall, almost five-hundred samples of canine MCTs, including some cell lines, have been investigated for somatic mutations of the *c-KIT* locus [33, 34, 36, 37, 43, 45, 50-54]. Hence, most of the activating *c-KIT* mutations have likely already been described, athough the rarer additional single nucleotide polymorphisms could not be excluded *a priori*. In particular, internal tandem duplications (ITDs), indels, insertions, deletions and single nucleotide substitutions have been reported. Many initial studies focused only on the ITDs of exon 11 which were initially described as most prevalent. This finding was further confirmed by a study which did not find any mutations at exons 16 through 20 [36]. Indeed, the ITDs of exon 11 represent the most prevalent *c-KIT* mutations accounting for approximately 65% of all mutations [45]. Nevertheless, recently, relatively frequent mutations in exons 8 and 9 were also found [45]. Overall, exon 8 may carry ITDs and a single base substitution in about 5 % of cases; exon 9 may carry two different single base substitutions in approximately 4% of cases and indel, insertions, deletions and ITDs in exon 11 in 17% of MCT cases. The most common mutation occurring in the kinase domain in human beings has been described in only 1 canine case [45]. Unfortunately, the indel mutation occurs at the splice site between exon 17-intron 17-18. The mutation has been found at the cDNA level and the genomic counterpart could not be ascertained (Dubreil P. personal communication). So far, there is only a single report describing an ITD involving exon 12 [33]. In cats, the role of *c-KIT* mutations in MCTs is fairly unclear and still under debate, although the activation of the ITDs of exon 8 has been described [38, 55, 56].

Much evidence exists that *c-KIT* mutations are correlated with increased tumor aggressiveness. *c-KIT* mutations are more frequent in grade II and III than in grade I [37, 50, 52, 57] which correlates with other prognostic markers, such as KIT aberrant cytoplasmic localization [37, 41, 42, 54]. Furthermore, Ki67 and AgNOR [54] are more frequent in those MCTs characterized by recurrence, metastasis or death [37, 50] and negatively correlate with disease-free interval and survival [57]. Assessment of *c-KIT* mutational status may also be pivotal in predicting treatment response to TKI. For example, Masitinib has been demonstrated to be particularly effective in those patients with mutated KIT, improving overall time-to-disease progression and overall survival, in particular, when used as a second line treatment [44]. Similar considerations could also be made for cats because *c-KIT* mutated neoplastic mast cell growth has been effectively inhibited by TKI *in vitro* [38].

2.3. Technical Issues

Evidence of the usefulness of the assessment of *c-KIT* mutation status and the resulting recommendation of its inclusion in the MCT work-up procedure deserves the effort of establishing and standardizing reliable assays. In addition to exon 11 ITDs, additional less common mutations have been described (see above). A comprehensive assay is usually expected to detect all possible mutations with unaffordable costs and long turnaround times. Conversely, the possibility of saving time and money using cDNA is impractical [34, 43, 45, 53]. Indeed, many different assays should be provided for the analysis of genomic samples. The possibility of analyzing genomic DNA would also allow the use of poorly conserved samples or archive specimens. The analysis of genomic DNA instead of mRNA allows the delivery of suitable samples to the reference laboratory.

This and other technical issues must be adequately resolved before *c-KIT* mutational status assessment can be considered a genuine routine molecular assay. Screening for *c-KIT* mutation has mainly been carried out using PCR followed by agarose gel electrophoresis for ITDs and direct sequencing for single base substitutions [34, 50]. In particular, canine *c-KIT* ITDs are reported to vary from 3 to 79 bp. Since agarose gel electrophoresis has low resolution and sensitivity, ITDs, or insertions and deletions smaller than 9-10 bp may likely be missed using standard agarose gel electrophoresis. In fact, fragment analysis carried out using polyacrylamide gel electrophoresis on a sequence analyzer has been shown to represent a more sensitive and reliable technique for ITD detection; the amplicons are obtained with fluorescent-labelled primers [52]. Fragment analysis is typically used in microsatellite analysis and is suitable for detecting even insertions and deletions as small as 1 bp. Furthermore, the simultaneous use of size standard and allelic ladder allows the exact measurement of the fragment (and consequently of the ITD) size without the need for direct sequencing. A further improvement is represented by GeneScanning carried out using capillary electrophoresis on an automated sequence analyzer. GeneScanning with capillary electrophoresis is very sensitive favouring the detection of rare alleles of different sizes with respect to wild type alleles, also when mixed with prevalent wild type alleles (personal observation, unpublished). In our group, we use a GeneScanning assay on an automated sequence analyzer which utilizes a fluorescent-labelled primer annealing on exon 11 and a reverse primer complementary to the intron 11 with an expected product length of 201 bp (Fig. **3**). PCR would also be able to detect a single ITD described, until now, as involving the 5'part of exon 12 [33]. At the genomic level, the mutation is believed to be represented by a duplication including the 3'end of exon 11, the entire intron 11 and the 5' end of exon 12; hence, a reverse primer complementary to the 3'end of intron 11 may anneal twice yielding 2 PCR products of different length using the same forward primer. Nevertheless, PCR should have an extension step lasting longer than necessary to allow the possible amplification of products of 600 bp or more. Furthermore, since the fragment analysis may typically detect products having a length up to 500-600 bp (also depending on the capillary used), longer fragments should be excluded by also running the product on agarose gel (Table **1**).

The issue of mutated alleles mixed with predominant wild type alleles arising from normal cells contaminating the specimen is both exciting and tricky. Indeed, the accumulation of somatic mutations propels oncogenesis from pre-neoplastic lesions to advanced highly metastatic cancer. In cancer biopsies, mutated alleles may be rare or underrepresented with respect to wild type alleles due to different causes. By themselves, clinical specimens are heterogeneous due to genetic evolution or the presence of normal or reactive cells surrounding the tumour. Screening methods include the amplification of the DNA by PCR followed by different downstream applications, such as direct sequencing among others. Nevertheless, PCR is inherently not selective for different alleles, and predominant alleles have a competitive advantage in the amplification process which leads to masking the presence of the mutated alleles in downstream applications. Moreover, the mutated allele is frequently heterozygous. This phenomenon is known but may have been underestimated. The ability of PCR and pyrosequencing to detect somatic mutations has been reported to be approximately 10%. This limit may require the laser capture microdissection (LCM) of the bioptic sample to enrich the sample of tumor cells. Nevertheless, LCM is not an inherently precise technique; it is cumbersome and requires expensive instrumentation. Furthermore, LCM may help to enrich the sample of tumor cells but may miss those cells with the additional rare allele.

A variant of conventional PCR promises to solve this problem with a very easy, cheap and intrinsically elegant method called co-amplification at lower denaturation temperature (COLD) PCR [58]. COLD-PCR is a new form of PCR which preferentially amplifies rare alleles from mixtures of wild-type and mutation-containing sequences, irrespective of where an unknown mutation lies. It will represent a cornerstone of somatic mutation detection in veterinary medicine allowing individualized molecular medicine and tumor genetic profiling in coming years. A detailed explanation of the method and protocols is not within the scope of this chapter and the reader is therefore referred to specific references for further details [58].

The method relies on the following observations: two amplicons differing by one single nucleotide have a 0.2-1.5°C difference in melting temperatures (Tm); each double strand DNA sequence has a critical denaturation temperature (Tc), which is lower than its Tm; PCR amplification efficiency for a DNA sequence falls abruptly

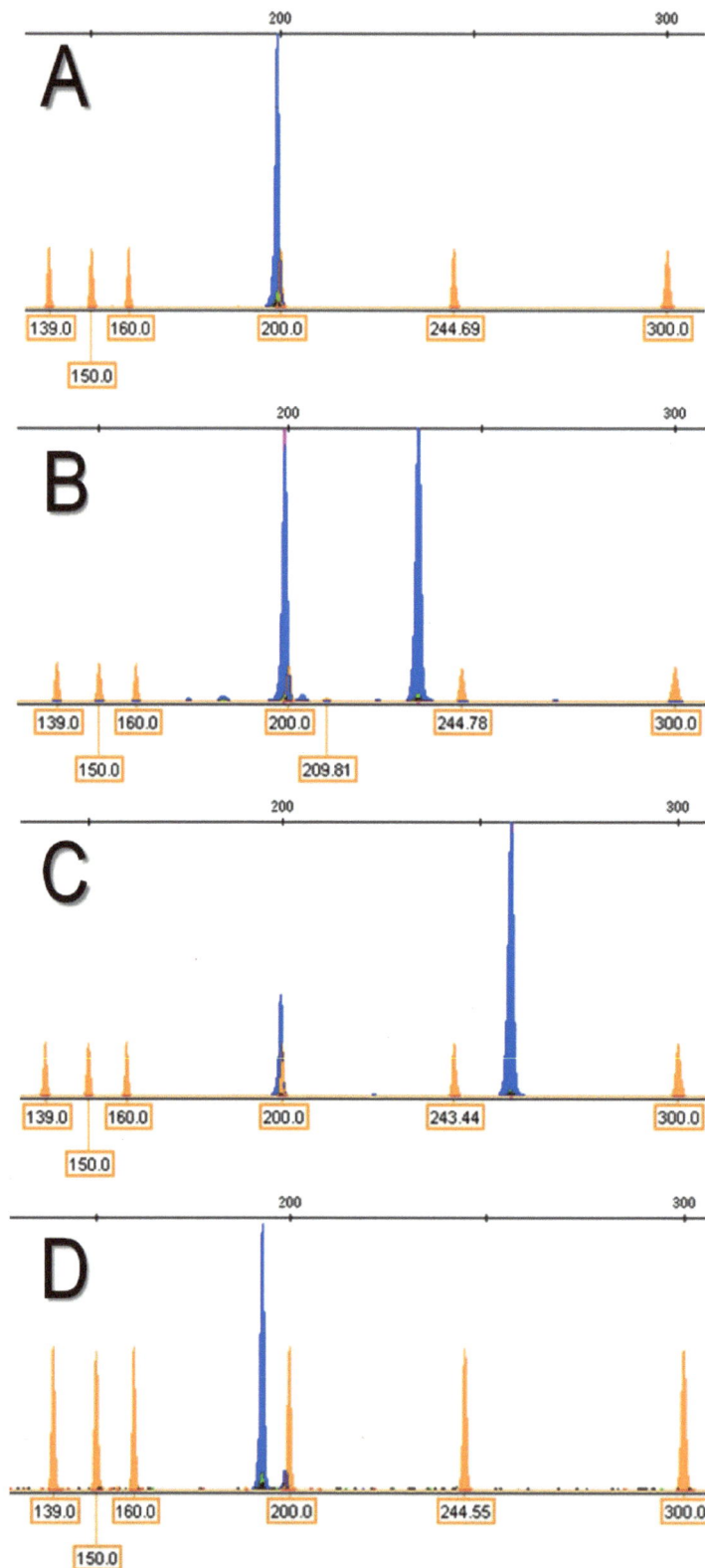

Figure 3: C-kit mutational status assay: GeneScanning findings of the exon 11 PCR assay. Selected examples of productive mutation within exon 11. A) wild type allele; B) ITD of 36 bp C) ITD of 54 bp and D) deletion of 6 bp. The orange peaks represent the size standard which indicates the apparent size of PCR fragment. The wild type allele, with an expected size of 201 bp, is consistently located together with the 200 bp size standard peak.

if the denaturation temperature is set below its Tc; DNA amplicons differing by a single nucleotide reproducibly have different amplification efficiencies when the PCR denaturation temperature is set to Tc [58]. These observations can be utilized during PCR amplification for selective enrichment of minority alleles differing by one or more nucleotides irrespective of their position. COLD-PCR exists in two forms: full COLD-PCR and fast COLD-PCR. Full COLD-PCR is used when the enrichment of all possible mutated alleles, including unknown alleles, is required. In full COLD-PCR, after a few cycles of conventional PCR, the cycle is modified by adding two steps. After denaturation at 94°C, a step lasting 7-8 minutes at an intermediate temperature of 70°C is added to allow the cross-formation of the mutant and wild type allele heterodimers; the temperature is then raised to the critical value which selectively denatures both the mutated homodimers and heterodimers and, finally, the cycle proceeds as usual lowering the temperature for primer annealing and then raising it for the extension step. The fast version is even simpler. If a Tm reducing mutation needs to be enriched, the COLD-PCR protocol is modified by denaturing at the Tc instead of 94°C. The alleles with a lower Tm are amplified more efficiently [58].

COLD-PCR followed by a downstream technique, such as direct sequencing or fragment analysis should be considered as the best option for the "*c-KIT* mutation detection" molecular assay. Indeed, this method enriches the minority alleles increasing the ability to detect the mutated alleles of neoplastic cells, even in a strong background of wild type alleles of normal surrounding cells without the use of LCM. Furthermore, unknown mutations may more readily be revealed.

To recapitulate, even if *c-KIT* ITDs of exon 11 account for the majority of somatic mutations in MCTs, about 35% of known mutations occur elsewhere in exon 8,9 and 17. Hence, the "*c-KIT* mutation detection" molecular assay should include further assays in addition to a PCR targeting exon 11, for example, a PCR amplifying exon 8 and exon 9. At least the one targeting exon 8 should employ a fluorescent-labelled primer suitable for fragment analysis and capable of detecting the reported 12bp ITD; the same PCR product should also be sequenced for known SNP identification; furthermore, direct sequencing should follow the PCR targeting exon 9. Ideally, COLD-PCR should be used as the preliminary step of the abovementioned assays. The COLD-PCR protocols for the amplification of *c-KIT* exons 8 and 9 currently used in our lab are reported in Table **1**.

Table 1: Protocols for *c-KIT* mutational status assessment

Target *c-KIT*	Primer 5' > 3'	PCR	Post PCR
exon 8	Hex - ggggagccttggtgaggtgt ccctgctgtccttccctcgt	25 ul of PCR reaction; DNA: 2.5 ul (up to 4 ul if the DNA was purified from formalin-fixed paraffin-embedded tissue); primer: 400nM each; Buffer AmpliTaq Gold™ 10X: 2.5 ul; MgCl₂ 25 mM: 1.75 ul; dNTPs 200 µM each: 2 ul; Ampli Taq gold™: 0.25 ul; water up to 25 ul Cold PCR program **1X:** 95°C for 9' **10X:** 95°C for 20'' - 58°C° for 20'' - 72°C for 20''	**Sequencing:** Load 5 ul of PCR product with 1 ul of loading dye in 1.5% agarose gel; electrophorese appropriately; check after staining with ethidium bromide for the presence of PCR product Purify the remaining PCR product with NucleoSpin Extract 2 (Macherey-Nagel); carry out the sequencing reaction in a final volume of 20 ul using the Big Dye® Terminator v1.1 kit; Ready reaction premix: 4 ul; Big Dye Sequencing Buffer: 2 ul; primer 3.2 pmol (forward and reverse in 2 separate reaction tubes);template 2 ul; HPLC quality water (Fluka®) to a final volume of 20 ul **GeneScanning**: Dilute the PCR product 1:20 and 1:40 with HPLC quality water (Fluka®). Mix thoroughly by careful pipetting 1 ul of diluted PCR product , 0.1 ul of Genescan ® 500 LIZ Size Standard (Applied Biosystems) and 20 ul of Hi-di Formamide ® (Applied Biosystems). Denature the PCR product by keeping it at 95°C for 5' and load the denatured diluted sample on the automated sequencer. The capillary shoul d be filled with POP4 polymer. Inject the sample at 15 Kv for 5 sec, and run at 60°C with Run Module: GS STR

		1X: 72°C for 10' **1X:** 95°C for 2' **40X:** 95°C for 20'' - 70°C for 8' – 83.4°C for 10'' – 58°C 20'' - 72°C for 20'' **1X:** 72°C for 10'	POP4(1ml)G5.md5. Always run the normal control in the same assay. Analyze both the raw data and the analyzed data using the following Software: Genemapper ID vers3.2 or peak scanner v1.1
exon 9	actcgtctctgtcaccgtctggaa atggcaggcagagcctaaacatcc	Same as for *c-KIT* exon 8 with the T$_c$ (in bold above) at **80.9°C**	**Sequencing:** as above
exon 11	6FAM - atgatctgtctctcttttctcccc gtacacaaaaaggttacatggaaagc	25 ul of PCR reaction; DNA: 2.5 ul (up tp 4 ul if the DNA was purified from formalin-fixed paraffin-embedded tissue); primer: 500nM each; Buffer AmpliTaq Gold™ 10X: 2.5 ul; MgCl$_2$ 25 mM: 2.5 ul; dNTPs 200 µM each: 2 ul; Ampli Taq gold™: 0.25 ul; water up to 25 ul **1X:** 95°C for 9' **40X:** 95°C for 25'' - 58°C°C for 25'' - 72°C for 45'' **1X:** 72°C for 30'	Load 5 ul of PCR product with 1 ul of loading dye in 1.5% agarose gel; electrophorese appropriately; examine after staining with ethidium bromide for product above 500 bp **GeneScanning**: as above

3. DIAGNOSTIC AND PROGNOSTIC POTENTIAL OF TELOMERASE ACTIVITY MEASUREMENT BY PCR

3.1. Introduction

One of the most popular applications of PCR in oncology is represented by the TRAP (Telomeric Repeat Amplification Protocol) assay for telomerase activity (TA) assessment. Since the original establishment of the methods by Kim *et al.* in 1994, many advances have greatly improved the accuracy of the assay. Such advances, mostly represented by improvements in primer design, clearly state the paradigm of the flexibility and robustness of PCR techniques.

3.2. Telomere Biology and Cancer

Chromosomes are discrete molecules composed of two linear strands of DNA. At each chromosome terminus, the DNA strands are truncated. Truncated DNA molecules are "sticky" and prone to interact with each other. At the chromosomal level, the sticky ends lead to cytogenetic aberrations, represented by chromosome instability, fusions and recombinations. To avoid the cells sensing the truncation as a double strand break and activating the DNA repair mechanism, a complex architecture of proteins (shelterin TRF1, TRF2, TIN2, Rap1, TPP1 and POT1) [59, 60] interacts with the terminal nucleotides of the chromosome stabilizing a T-loop nucleotide structure which hampers a double strand break (DSB) recognition and chromosomal aberration events. Such a chromosome endcapping structure is called a telomere. The telomere is thus composed of very long arrays of short tandem repeats arranged in double-stranded DNA ending with a terminal single-stranded 3' overhang. In vertebrates such tandem repeats are well conserved among species and represented by the TTAGGG sequence. The 3' overhang is folded and stabilized by shelterin proteins to form the T-loops (Fig. **4A**).

Due to the mechanism of DNA replication in the 5' to 3' direction, the leading strand is completely replicated. Conversely, on the lagging strand, the RNA priming does not allow for complete duplication.

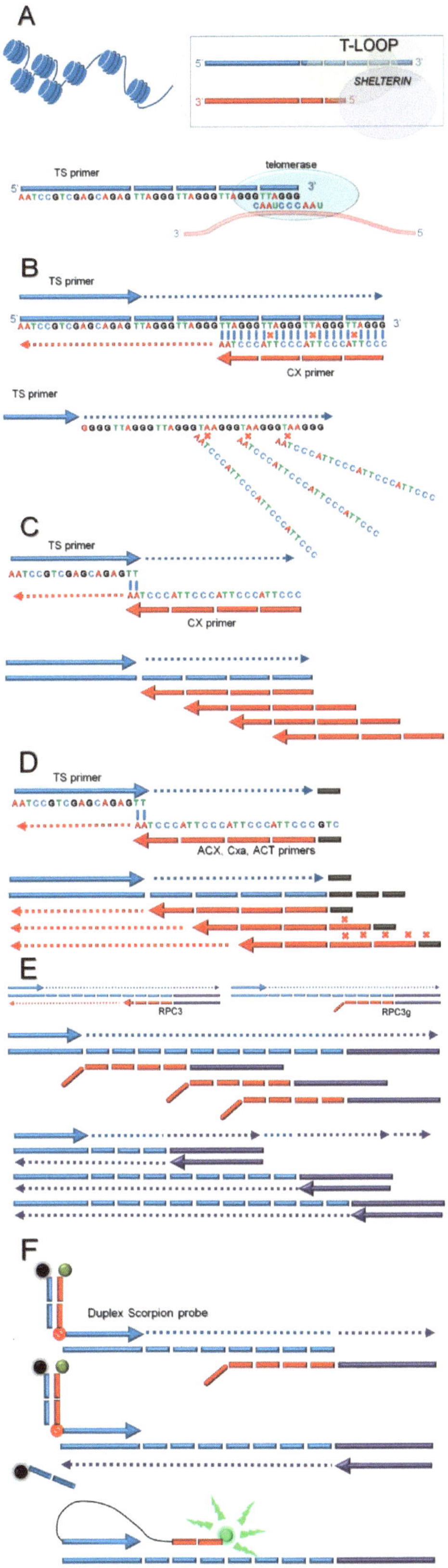

Figure 4: Telomerase assay (TRAP); Schematic representation of different TRAP assays.

This leads to the shortening of the lagging strand which loses many terminal TTAGGG repeats after each replication cycle. It has been calculated that approximately 30-200 bp of telomere sequences are lost at each replication cycle [61]. When the attrition of the telomere invariably reaches a critical point, most DNA binding sites are lost and the T-loops are destabilized. Thus, DNA is exposed to the DSB sensors which activate the pro-apoptotic pathways leading to cell senescence and death [62].

Telomere length differs widely among species. In humans, telomeres range between 10-15 kbp [62] while rodents have much longer telomeres (25-100 kbp), [63]. It has been proposed that such a difference between humans and rodents hampers the possibility of using mice and rats in biomedical research on telomere biology. Alternatively, dogs could represent a reliable animal model [64]. Indeed, canine telomere lengths are vastly heterogeneous but comparable with humans. An approximate canine telomere length ranging from 3 to 23 kilobases was found [64-67]. Similar values were also found in feline and equine telomeres [68, 69]. Nevertheless, unlike humans, the shortening of telomeres along with increasing age was suspected but has not yet been clearly demonstrated in dogs [66]. Significant canine breed differences were advocated to explain this finding [66]. Moreover, attempts to identify tissue length specificity were inconclusive as well as any differences between normal and neoplastic tissue [65]. In human beings, telomeres in cancer are stabilized at a shorter length. Similar results have not been confirmed in dogs with the exception of cell cultures [70].

In humans, all somatic cells undergo telomere attrition with a few notable exceptions represented by immortal stem cells, activated lymphocytes and germ cells. Conversely, it has universally been established that telomere maintenance above the critical threshold of T-loop de-stabilization is a hallmark of almost all neoplastic cells. Although alternative pathways for maintaining telomere length were found [71], until now, the main mechanism for avoiding the critical telomere consumption in cancer cells is controlled by the activity of the enzyme, telomerase. Originally, telomerase activity (TA) was found in almost 90 % of cancer cells and none was found in normal adult somatic cells [72]. Over a decade ago, this exciting breakthrough led many to postulate a new era in both cancer diagnostics and therapy. Unfortunately, after many years the telomerase road to defeating cancer is becoming increasingly complicated and much of the enthusiasm has faded. Nevertheless, the potential of TA as a diagnostic and prognostic target in support of conventional techniques in molecular diagnosis is well established.

Telomerase holoenzyme is composed of several subunits including a catalytic subunit with reverse transcriptase activity (TERT), an RNA component (TERC or TR) used as a template to add telomeric tandem repeats and other proteins; telomerase associated protein 1(TEP1), Heat shock protein 90(Hsp90), p23 and dyskerin [73]. In dogs, TERT and its promoter have been cloned and characterised [74, 75]. TERT expression is restricted to those cells with TA, although very few copies per cell have been demonstrated. On the other hand, the RNA component is highly expressed ubiquitously regardless of TA, although the RT expression is five-fold higher in tumour cells [76, 77]. Although TERT and RT constitute the basic apparatus for TA *in vivo*, other telomerase- associated proteins, such as the shelterin complex, are necessary for enzymatic activity. Additionally, telomerase has other functions besides that of telomere lengthening associated with the induction of cancer, such as the ability of conferring apoptosis resistance [78].

3.3. Telomerase Activity Measurement Using PCR-Based Methods

Due to the extremely low copy number of the telomerase holoenzyme (1-5 /cell), an extremely sensitive tool is necessary for the detection of its enzymatic activity. To achieve the appropriate sensitivity, a PCR-based method was developed by Kim and colleagues in 1994. The telomeric repeat amplification protocol (TRAP) assay is an all in one 2-step *in vitro* reaction which utilizes the ability of the functioning telomerase to extend an artificial primer called telomerase substrate (TS) with TTAGGG repeats (Fig. **4A**). In the subsequent step, the extended telomeric repeat is PCR-amplified using the same TF as the forward primer and a primer annealing on the telomeric repeat as the reverse primer (RP). The PCR products are visualized, after gel electrophoresis, as a typical ladder of 6 bp amplicons.

The assay requires the extraction of the intact holoenzyme from the tissue. This step is critical since the RNA component should not be degraded by RNAses. To maintain TA, snap freezing in liquid nitrogen and storage at -80°C are recommended. The tissue extracts are incubated at 30°C for a period ranging from 10 to 60 minutes in a reaction mixture containing the tissue extract, a reaction buffer (20 mM Tris-HCl (pH 8.3), 1.5 mM $MgCl_2$, 63 mMKCl, 10 mM EGTA, 0.2 mg of T4 gene protein), the TS primer, dNTPs, the RP and Taq polymerase [72].

The original assay had reliable specificity and sensitivity but also had some drawbacks: 1) it was qualitative, 2) the detection methods included hazardous radioactive labelling and time-consuming visualization by autoradiography after gel electrophoresis, and 3) some false positive results were possible due to primer-dimer elongation with a typical hexamer pattern (Fig. **4B** and **4C**).

To overcome all the above-mentioned drawbacks of the original TRAP assay, many modifications and improvements have subsequently been described. In the original assay, the RP called CX (5'-CCCTTA)₃CCCTAA -3') was partially T:T mismatched to limit the staggered annealing of the primer on the telomeric repeats leading to an artefactual elongation of the amplicons during the PCR step [79] (Fig. **4B**). Another drawback was represented by the annealing of the CX primer on the TS primer, yielding primer-dimer artefacts [79] (Fig. **4C**). Both occurrences could lead to artefactual elongation during PCR, affecting the accuracy of the original assay. Different modified RPs have been described for reducing staggered annealing. In particular, CXa and ACX reverse primers have an additional 5' tail of nucleotides ("anchor") not complementary to the telomeric sequences. The anchored primers prevent further extension of both the telomeric product and the primer-dimers, impeding the staggered annealing of the RP in the 5' terminal anchor sequence and its elongation [79, 80] (Fig. **4D**). Further improvements have also been achieved in the TP-TRAP assay. A 5' extended TS (MTS) primer as well as two distinct reverse primers (RP and RPC3) have been utilized [81]. The MTS primer allows for an increase in annealing temperature while the RPC3 primers have a 3' end complementary to 3 telomeric repeats and a 5' long anchor not complementary to the telomeric sequence. The RP primer anneals with the 5' anchor of the RPC3 primer (Fig. **4E**). The 2 reverse primer strategy initially involves a 2 cycle PCR at a low (50°C) annealing temperature to allow the annealing of the 3' end telomeric specific RPC3 primer. After that, the increase in the annealing temperature to 63°C in an additional 25 PCR cycles allows the RP primer to anneal within the RPC3 sequences included in the amplified products. This strategy completely prevents the possibility of staggering annealing and artefactual product amplifications [81] (Fig. **4E**). A further modification of the TP-TRAP assay has been reported by including a GG dinucleotide at the 3' end of the RPC3 primer (RPC3g). The oligos hybridize to the telomeric sequence during the first 2 cycles of the PCR but the 3' end could not be extended whereas its 5' anchor sequence is included in the amplicons and targeted by the RP primer in the cycles from 3 to 27 [82] (Fig. **4C**). Besides much improved specificity, the modified TP-TRAP assay conferred an extended linearity to the assay and allowed the precise measurement of the TA. Initially, the quantification method used a 32P radiolabelled MTS primer [81, 82] which limited its popularity. The establishment of a PCR-ELISA kit (teloTAGGG PCR ELISA plus kit (Roche molecular diagnostic) and its commercial availability popularized TA measurement. The ELISA-PCR uses a TS primer with a biotin moiety at the 5' end. The PCR amplified products are immobilized *via* the binding of streptavidin-coated plate wells. A further hybridization DIG-labelled probe is detected using a specific antibody conjugated with horseradish peroxidase and its respective substrate TMB. The amount of TA is measured as a comparison to an internal control template in order to improve linearity and assess the presence of PCR inhibitors.

In veterinary medicine, an alternative conventional assay which utilizes gel staining with SYBR® Green I has become popular (TRAPeze™ Telomerase detection Kit, Oncor). Nevertheless, all the conventional TRAP methods have inherent problems which are represented by the quantification of PCR end-point products with an inevitably restricted quantitative dynamic range.

3.4. Advances in Telomerase Activity Measurement by Quantitative Real Time PCR-Based Methods

To reduce time consuming post-PCR steps and improve quantification, many methods of real time PCR have been validated [83-86]. Real-time monitoring allows the quantification of PCR product accumulation during the exponential phase. The dynamic range is consequently markedly expanded. Initially, a RQ-TRAP assay using SYBR® Green I showed a much improved dynamic range and sensitivity when compared to conventional TRAP [83]. The method included only the ACX reverse primer thus some problems of primer-dimer artefacts were not completely resolved. This method revealed that the TA half life was 5-11 hours (50% less than previously thought) and that TA was much lower than demonstrated with conventional TRAP. The method was further simplified and optimized by eliminating some reagents (T4 gene protein and EGTA) from the reaction mixture and reducing the primer-dimer by refining the annealing temperature of the PCR.

A fluorescent-based assay relying on the Amplifluor ® primer was demonstrated to achieve increased sensitivity. In the TRAPeze™ XL, the conventional kit was modified with a reverse primer modified with 5' additional sequences having a hairpin loop. A fluorescent dye (FAM or TAMRA) at the 5' end and a quencher molecule (DABSYL) linked to the nucleotide are in close proximity to the fluorophore in the free primer hairpin structure. During DNA polymerization, the Amplifluor ® primer is displaced, generating a fluorescent signal proportional to the amplification product [84].

A refined and elegant method known as DS/TP-TRAP assay, which has improved the specificity of the previous RTQ-TRAP assay has recently been described. The method includes a Scorpion probe as a TS forward primer and the double reverse primer (RP and RPC3g) of the modified TP-TRAP method [86].

The method is inherently robust, sensitive and specific. Overall, the DS/TP-TRAP assay takes 3 hours vs. the 24h of the original TRAP assay; it has a linearity spanning from 1 to 100000 cells *vs.* 250 – 5000 cells of the original TRAP assay.

The assay relies on duplex Scorpion probes as a forward primer instead of the telomerase substrate. The duplex scorpion is composed of the scorpion probe and a quencher oligonucleotide. The Scorpion probe is made up as follows from 5' to 3': a fluorophore, the telomeric-specific sequence, a PCR blocker and the sequences used by the telomerase to add telomeric repeats. The quencher oligonucleotide is composed of a sequence complementary to the telomeric specific sequences and a quencher molecule at the 3' end. In the reaction mixture, the quencher oligonucleotide anneals with the 5' telomeric sequences of the scorpion probe and the quencher falls in the proximity of the fluorophore quenching the fluorescence (Fig. **4F**).

To the best knowledge of the authors, the recent advances described above have not yet been utilized in veterinary investigations of telomere cancer biology. Certainly, many of the shortcomings of the initial TRAP assays may have been overcome by these novel techniques and, hopefully, TA may become even more popular in the clinical settings.

3.5. Usefulness of Telomerase Activity Measurement in Veterinary Medicine

Much evidence has accumulated indicating that pets in general, and dogs in particular, could represent an appropriate animal model for telomerase-targeted therapy since there are overlapping aspects of molecular biology, histopathology and response to therapy. Furthermore, telomerase is a widely diffused cancer associated antigen having potential prognostic and diagnostic implications.

A commercially available TRAP assay has also made the investigation of TA possible in veterinary medicine. Until now, telomerase has been investigated in a multitude of animal cancers. Overall, more than 90% of cancers in pets show TA while no such activity have been found in the vast majority of normal tissues [64, 65, 87, 88]. Including TA assessment as a screening tool for malignancies in cats has also been proposed [89].

The potentiality of TA measurement as a diagnostic tool to distinguish neoplastic from non-neoplastic effusion, which continues to represent a challenge for the clinical pathologist, has also been explored. However, a TRAP assay performed worse than cytology with low specificity and sensitivity [90] and confirmed previous unrewarding studies in humans. TERT immunoreactivity in canine brain tumors correlated with the WHO histological grade and Ki67 immunoreactivity [75]. In mammary gland tumours contradictory findings were reported. Initially, TA was found both in benign and malignant neoplastic lesions but not in hyperplastic or normal tissue. In mammary malignant tumors, TA activity was significantly higher than in benign ones [87, 88]. However, in another study, TA correlated with TERT immunoreactivity, but both were also found in non-neoplastic lesions and in normal mammary tissue [91]. TA was also demonstrated in normal canine lens epithelial cells [92], lymph nodes [93-95], myocytes [96] and testes [87]. TA correlated with well-established histopathological prognostic indicators, such as PCNA antigens but not with tumor diameter, metastasis and recurrence [88]. 75% of canine osteosarcomas have TA, although the presence of TA did not predict either disease-free intervals or survival times [97].

In cats, 29 out of 31 solid neoplasias of different types and only 1 out of 22 non-neoplastic samples were shown to possess TA with a sensitivity and specificity of 94% and 95%, respectively [89]. Both telomerase expression and TA showed a significant and complementary discriminative ability to detect malignancies in feline mammary tumors [98]. Indeed, although TA measurement was thought to be a milestone in the diagnosis of cancer, many pitfalls were found and more recent studies have raised doubts about its usefulness as a prognostic and even a diagnostic tool [99].

These limits and pitfalls are most likely due to the presence of TA in normal cells and to the technical limits of the conventional kits used in veterinary studies. It has been shown that alternative pathways (ALT) involving recombination and DNA repair mechanisms are involved in telomere maintenance and cell immortalization without TA. Obviously, in such a context, TA does not have diagnostic and prognostic potential. Even if it is becoming evident that improved quantitative RQ-TRAP could overcome the drawbacks correlated with the detection of TA in normal cells, RQ-TRAP has not yet been employed for assessing TA in pets. Many of the above-mentioned contradictory results will likely be clarified by its use.

4. PHARMACOGENOMICS OF P-GLYCOPROTEIN

ABCB1-1Δ (ATP Binding Cassette subfamily B, member 1; formerly Multi-Drug Resistance *MDR1*) is the gene coding for p-glycoprotein, a drug efflux member pump which transports xenobiotics from within to outside the cells. *ABCB1-1Δ* is expressed in numerous normal tissues. In particular, p-glycoprotein is responsible for the selective permeability of blood brain barriers. In veterinary medicine, the interest in *ABCB1-1Δ* depends on the possibility that germline mutation affecting the gene is responsible for a severe adverse reaction following the administration of drugs, such as macrocyclic lactones (ivermectin, milbemycin oxime) [100] and Vinca alkaloids [101]. The latter, including Vincristine and Vinblastine, are among the drugs most used in cancer chemotherapy in pets. It has been demonstrated that the animals carrying a 4bp deletion in the *ABCB1-1Δ* gene show a dose-limiting hematological toxicity, characterized by marked neutropenia and thrombocytopenia, following the administration of vincristine [101]. The effect is also evident, though to a lesser extent, in heterozygote dogs. The mutated allele is highly prevalent in some canine breeds such as the Collie (33% of homozygotes mutated and 43 % of heterozygotes), and in related breeds, such as the Shetland sheepdog, Australian shepherd, wäller, old English sheepdog and border collie [102], white Swiss shepherd [103] and German shepherd [104]. PCR-based genetic testing for the mutation responsible requires visualization on either high resolution polyacrylamide gels or even better with fragment analysis after capillary electrophoresis. It represents the first example of a pharmacogenomics assay useful for the management of chemotherapy in dogs.

5. FUTURE PERSPECTIVES

5.1. Canine MHC Diversity and Bone Marrow Transplantation

Allogenic transplantation of either bone marrow or umbilical cord blood (UCB) cells would be feasible opportunities to dramatically improve our ability of eventually achieving a complete cure in lymphoproliferative disorders of pets [105]. Furthermore, and unlike solid organ transplantation, in blood malignancies, the donor would not experience detrimental damage to its health and, thus, ethical issues would not be a concern.

On the other hand, there are still some limitations which prevent the diffusion of allogenic transplantation in pets. Allogenic transplantation is possible between Major Histocompatibility Complex (MHC) I and II matched donors and recipients. Initially, MHC matching was performed using serologic (microcytotoxicity assay) and cellular (mixed lymphocyte culture assay) methods. These techniques are quite laborious, time-consuming and sometimes inaccurate. More recently, the rapid and accurate molecular typing of extremely polymorphous MHC I and II loci has emerged. Molecular typing is performed using PCR and a downstream application, such as single strand conformation polymorphism (SSCP) analysis, restriction length fragment polymorphism (RLFP) or direct sequencing, accurately and rapidly [106, 107]. The MHC complex is called Dog Leukocyte Antigen (DLA) system in the canine species (and FLA in the feline species and so on in other species). DLA allele information has dramatically increased in the last decade [108-112] and public databases have been accumulating and organizing all the information as freely available online resources (http://www.ebi.ac.uk/ipd/mhc/). A class I and II molecular-based assay in dogs has been developed by assessing the alleles for class II loci by sequencing only DLA-DQB1 exon 2, and for class I loci DLA-88 exons 2 and 3 [107, 111, 113]. After PCR, allele typing could easily be assigned by aligning the sequenced loci with the allele sequences of databases.

DLA typing for donor/recipient matching is pivotal for allowing early engraftment and avoiding graft versus host disease (GVHD). The availability of MHC-matched siblings would be feasible in veterinary medicine while the availability of bio-banks with unrelated DLA-typed donors would be less practical. UCB stem cells seem to circumvent some of these pitfalls. Indeed, UCB transplants are more tolerant to partial MHC mismatches leading to less frequent and severe GVHD [114]. The main limitation of UCB is represented by later engraftment due to less HSC grafted in the donor. Nevertheless, MHC-typed HSC could be expanded *in vitro* for reliable clinical use. Although of lesser stringency, also in this case, MHC typing is relevant for successful engraftment (reviewed in [114, 115]).

Another challenging problem of allotransplantation in veterinary medicine was represented by the amelioration and refinement of the conditioning regimen used to treat the recipient in order to allow complete engraftment. Some of the protocols, now used in human medicine were, howver, established in canine models. Thus, their use is a prime example of translational medicine. For instance, non-myeloablative total body irradiation before the transplantation and a short term course of postgrafting immunosuppressive therapy with mycophenolate mofetil and cyclosporine [116, 117] has proven effective in DLA-matched sibling marrow transplantation and would be practical in some reference veterinary hospitals.

5.2. High-Throughput Analysis of Gene Expression in Cancer – Gene Profiling

In most respects, cancer is inherently a heterogeneous disease and very complex cell functions are simultaneously altered. The reductive idea that studies of the expression of a few genes at a time could disclose those underlying mechanisms and clarify the pathogenesis of the cancer or even be translated into useful markers has been abandoned in cancer research for some years. Based on human and experimental animal findings, some studies on a few selected genes have also been carried out in tumors found in domestic animals [99, 118-124]. Many of these studies have contributed to the establishment of animal models or the confirmation of some pathway alterations but, unfortunately, from the clinical standpoint, the potential prognostic or diagnostic importance of their findings is uncertain and the impact on clinical decision-making is limited. Conversely, the term "Gene expression profiles" (GEPs) stands for methods which simultaneously assess the large-scale relative abundance of a multitude of mRNA in diseased tissues as compared to the respective healthy tissues. The mRNA expression profiles of selected neoplasias represent their specific "signature". GEPs have been implemented thanks to the advent of DNA array technology. Microarray and complex clustering algorithms have greatly increased our understanding of cancer biology, have revolutionized our traditional categorization systems and have allowed the establishment of many molecular assays with exceedingly reliable diagnostic, prognostic and predictive information (reviewed in [125, 126]).

PCR is an ancillary of DNA microarray technology. First, DNA microarray chips can be purchased or can be made by the individual laboratory, spotting PCR products on the chip surface. Second, real-time PCR is used to overcome some drawbacks of DNA microarray technology; RT-PCR is much more inexpensive and more accurate in quantifying, and the results are more reproducible and easier to interpret than complex chip output. Typically, the expression of the most meaningful genes as ascertained by DNA array is included in RT-PCR mRNA expression profiles. For instance, canine mammary cell lines have been investigated using DNA microarray profiling. DNA microarrays have shown upregulation in most of the genes clustered in the Wnt signaling, cell cycle regulation, complement-signalling, integrin signalling, and cytoskeletal and Rho-GTPase signalling. Those findings were further confirmed and validated using RT-qPCR of only 13 genes which were representative of all the abnormal pathways [127]. Similar studies have also recently been carried out on spontaneous canine mammary [128] and brain tumors [129]. Therefore, it is most likely that the notable impact of the mole of data obtained with these high-throughput techniques is also forthcoming in pet medicine oncology, and large validation studies using RT-qPCR will be needed as soon as possible.

5.3. Genetic Predisposition to Cancer-Disease Susceptibility Loci

Cancer is a disease caused by exposition to environmental factors, such as waste pollutants, radiation, UVA/B, cigarette smoke, viral agents and diet, all interacting with the genetic background. Environmental factors induce genetic and epigenetic mutations leading to cancerogenesis since it can be claimed that "cancer is a genetic disease which only rarely is also heritable".

About 70% of tumours affecting humans are sporadic and correlated with the above-mentioned risk exposure leading to acquired somatic mutations or epigenetic events. In the remaining 30% of cases, germline mutations are responsible for cancer susceptibility with, therefore, an inherited familial genetic background. Cancer predisposition syndromes may be due either to low (20-25%) or high (5-10%) penetrant alleles (Reviewed in [130]).

The comprehensive record of germline mutations responsible for cancer susceptibility in humans is broad. A selected list of those genes is reported in Table **2**. Due to the impressive amount of information available, cancer genetic counselling is becoming a leading specialization of genetic counsellors who have to manage people who are at increased risk of developing cancer [131].

The predisposition to cancer of various animal breeds has been widely recognized. Apart from anecdotal reports, much scientific evidence has accumulated on the tendency of certain canine breeds to develop cancer or even particular cancer types (Boxers, Labrador Retrievers, Bernese Mountain dogs, Scottish Terriers and Rottweilers) and, unlike humans, in veterinary medicine, the breeding selection could conceivably be used to eradicate cancer-suspectible genes.

Nevertheless, with the sole exception of PCR genetic testing for naturally occurring inherited canine renal cystadenocarcinoma and nodular dermatofibrosis in German Shepherd Dogs [132], there is no detailed knowledge of cancer predisposition mutations. Research is required to investigate the genetic cause of cancer predisposition. Population-genetic based studies have been initiated and are still ongoing. So far, there have been some "clues" of germline mutations in candidate cancer genes, such as, for instance, *p53*, *MET* and *ATM* [49, 133-136]. Hopefully, in the near future, more PCR testing will be available to assist breeders in limiting the frequency of those alleles involved in cancer susceptibility.

5.4. Genomic Instability and Cancer Prognosis

5.4.1. Loss of Heterozygosity

According to the multi-step model of tumorigenesis, neoplastic transformation and progression towards more malignant phenotypes are caused by subsequent acquired genetic mutations (somatic mutations). Somatic mutations may cause an increase in oncogene functions or disrupt tumor suppressor genes (TSGs) leading to the loss of functions concerning cycle control, apoptosis, cell growth and differentiation. Overall, the cumulative genetic events which accumulate during cancer development are referred to as genetic instability. Two genetic instability mechanisms occur in cancer: chromosomal instability (CIN) and microsatellite instability (MSI). One of the most likely consequences of general CIN in advanced cancer is the loss of heterozigosity (LOH) at multiple loci. The term LOH refers to the loss of genetic information of one of the two germline heterozygous alleles at a gene locus. When LOH occurs at loci bearing TSGs and leads to the loss of the sole functioning allele, the event might represent the second "hit" and be correlated with tumorigenesis. The most common mechanisms of LOH are, among others, deletion, gene conversion, and mitotic non-disjunction or recombination mediated, while the first "hit" had been an intragenic single nucleotide mutation leading to the first TSG allele inactivation (reviewed in [137]).

Loss of heterozigosity can be detected using different methods; the most common is represented by a PCR of genetic markers as a microsatellite (see below) followed by polyacrylamide or capillary electrophoresis. Other techniques described include, kariotype analysis, FISH, PCR+RFLP, SSCP, and other less common techniques. The analyses utilize the comparison between the same PCR amplification products of many different loci from neoplastic tissue and from corresponding normal healthy surrounding tissue (germline alleles). Only those markers which are heterozygous at the germline level can obviously be considered indicative of LOH markers for each patient. Some markers are so close to fundamental TSGs that their loss clearly parallels the loss of one TSGs allele.

Even if LOH is quite common in advanced cancer involving up to 50% of the chromosomes, it is also rarely seen as an early event of cancerogenesis. Sometimes, the presence of early LOH may signal an increased risk of progression from an oral dysplastic lesion to cancer (reviewed in [138]). Conversely, the LOH of the APC gene which is involved in cell cycle regulation has been detected in almost 90% of colorectal cancers (CRCs), likely representing the model where the molecular defects leading to cancerogenesis are best known. In CRC models of tumorigenesis, APC LOH is of no value in stratifying patients and, therefore, its prognostic and predictive importance is null [139].

By reviewing the literature of human CRCs, it emerges that studies on molecular markers and LOH, in particular, frequently yield contradictory results. As we might expect, meta-analysis reviews are necessary for establishing the role of those markers in their clinical context. Thus, once again, the finding and thorough validation of LOH markers in large studies are required prior to their introduction in clinical practice. Therefore, even though LOH is conceptually straightforward and its detection is not difficult; LOH studies in veterinary medicine are still very rare. Herein, two applications in veterinary medicine will

Table 2: Selected cancer susceptibility syndrome in human beings (modified from Garber *et al.* 2005)

Gene	Syndrome (OMIM entry)	Mode of Inheritance
BRCA – 1/ BRCA - 2	Hereditary breast and ovarian cancer –prostate- (113705, 6001859)	dominant
BRCA - 2	Fanconi anemia, medulloblastoma (6001859)	recessive
RET	Papillary renal cancer, Multiple Endocrine neoplasia (MEN) 2 Medullary thyroid cancers (171400)	dominant
MET	Papillary renal cancer syndome (605074)	dominant
PTEN	Cowden syndrome-endometrial, thyroid, breast cancer – (158350)	dominant
KIT	Gastrointestinal stromal tumors (606764)	dominant
P53	Li-Fraumeni syndrome (151623)	dominant
ATM	Ataxia-Telangectasia syndrome -mammary tumors, lymphoma, leukemia –(208900)	recessive
RB1	Retino blastoma (180200)	dominant
MEN	Multiple endocrine Neoplasia (MEN) 1 - (131100)	dominant
WT 1	Wilm's tumors	dominant
P16/CDK4	Hereditary melanoma	dominant

MLH1/MSH2/MSH6	Colon cancer Dominant MLH1, endometrial cancer MSH2, ovarian cancer MSH6, renal pelvis cancers, ureteral cancers, pancreatic cancer, stomach and small bowel cancers, hepatobiliary cancers	dominant
APC	Familial polyposis (colon cancer – (175100	dominant
VHL	von Hippel-Lindau syndrome -Hemangioblastomas of retina and central nervous system Renal cell cancer Pheochromocytomas- (193300	dominant

be mentioned: LOH was advocated as a pivotal mechanism of folliculin gene inactivation (2nd hit) leading to the neoplastic transformation of renal and cutaneous tissue in renal cystoadenocarcinoma and nodular dermatofibrosis of German Shepherd Dogs, a recessive trait disease due to germline mutation in the folliculin gene [132, 140]. The studies gave a very interesting insight into the pathogenesis of this model of the Birt-Hogg-Dube' syndrome, an inherited renal cancer using LOH as an elegant research tool. In the second example, investigators found some polymorphic markers, namely three microsatellite mutations and an ins/del mutation within the canine BRCA2 gene. These markers were shown to be polymorphic at the germline level and LOH was demonstrated in mammary tumors of dogs [141, 142]. These studies have demonstrated that LOH analysis is just at its beginning. Hopefully, additional markers and systematic validation in context will eventually disclose their potential prognostic importance.

5.4.2. Microsatellite Instability

Microsatellites are repetitive units of 1-6bp in the non-coding (the vast majority) and the coding (much rarer) regions of chromosomes. The most common microsatellites are represented by di and tri-nucleotides repeated in a tandem array. Microsatellites are highly polymorphous since many different alleles, usually represented by a various number of repetitive units are displayed in a given population. Although the definitive explanation of the origin has never been clarified, it has been postulated that polymerase slippage during replication may occur due to the looping out of either the replicating (1 more repetitive unit) or the template strand (1 less repetitive unit). Besides their biological role, which is under debate, microsatellites are used in several genetic applications: genome mapping, forensic and molecular epidemiology, *etc.* MSI, that is the presence of different alleles between the germline (healthy tissue) and the neoplastic tissue, is considered to be a feature of more complex genomic instability. MSI may be due to both genetically transmitted germline mutations or acquired silencing occurring in the gene involved in the DNA mismatch repair system (MMR). In some cases, MSI may be strictly associated with tumorigenesis and the investigation of microsatellites may clarify its pathogenesis (triplet expansion diseases). In other cases, MSI is a hallmark of genomic instability and may have a prognostic value. Nevertheless, the role of MSI as a marker is quite complex; indeed, in endometrial carcinoma, MSI is positively correlated with grade whereas in colorectal cancer MSI has been demonstrated to have a better prognosis [143, 144]. Preliminary findings have also recently been reported in veterinary medicine. High level MSI has been reported to occur in canine mammary tumors although its relevance in tumorigenesis remains to be elucidated [145].

5.5. Epigenomic Instability: PCR for Assessing the Methilome in Cancer

Since TSGs act in a recessive manner in tumorigenesis, the two alleles should both be inactivated in order to play a role in cancer progression. Besides intragenic mutations, another mechanism of inactivation is represented by hypermethylation of the CpG reach promoter region of TSGs which induces the transcriptional silencing of the downstream gene (reviewed in [146, 147, 148]). Many other abnormal DNA methylation events, such as global hypomethylation *etc.*, are known to influence the development and progression of cancer, generally accounting for epigenetic instability.

Although studies on epigenetic instability in veterinary medicine, with the notable exception of laboratory animals, are still embryonal [149-151] and far from clinical applications, the impressive amount of data regarding human cancer pathogenesis, the possibility of "diagnostic signature assessment based on the methylation pattern" and the evidence that aberrant DNA methylation is an early occurrence in cancer development have convinced the authors to include a brief overview of PCR techniques. DNA methylation studies have both the possibility of further supporting the translational medicine of companion animals and of becoming routine molecular assays used by veterinary practitioners.

For the most part, DNA methylation occurs by adding a methyl residue at the 5' carbon of the pyrimidine ring of the cytosine to form 5methylcytosine (5meC). In particular, the methylated cytosine is found in a 5'-CG-3' palindromic sequence called CpG. CpG reach regions are frequently found in the CpG island at the promoter or other regulatory regions of the DNA. The methylation of the promoter region prevents the transcription of the downstream loci. When the downstream loci code for a TSG, the aberrant methylation may lead to gene silencing and cancer development. As stated above, either the evaluation of the global amount of methylation (5meC) in the entire genome or the assessment of methylation in particular sequences, such as the CpG island, are relevant in cancer medicine. The former are routinely assessed using non-PCR-based methods; therefore, they are not mentioned herein. Conversely, PCRs are pivotal in many of those techniques used to assess 5meC in specific sequences. The cornerstone of these techniques was the DNA bisulfite-conversion reaction which uses sodium bisulfite to convert the unmethylated cytosine to uracil, ultimately incorporated as thymine during PCR, while leaving the 5meC unaltered. [152]. Starting from the bisulfite-treated DNA, a multitude of techniques have been developed to assess the methylation state, exploiting the differences evidenced between bisulfite-treated (5meC and U or T in place of unmethylated C) and untreated DNA. Many of those techniques rely on PCR (reviewed in [153]).

Methylation-specific PCR (MSP) was first described in 1996 [154]. MSP relies on the PCR amplification of CpG reach sequences (especially promoter regions). Since methylation primarily occurs at the CpG site, those cytosines outside a 5' – CG – 3' sequence are unmethylated and become U(T) after the bisulphite reaction. Exploiting such a phenomenon, three primer pairs have to be designed one targeting the untreated DNA, one targeting the sequences of bisulphate-treated DNA with T instead of C (unmethylated cytosine), and the last targeting the sequences of bisulphate-treated DNA with methylated cytosine. It should be taken into account that, after bisulphite treatment, the double strands of DNA are no longer complementary; therefore a primer may be designed in either a positive or a negative strand. To increase methylation specificity, the critical 3' end of each primer should ideally anneal within a CpG site to facilitate the sequence differences induced by bisulphite treatment. It is notable that MSPs are exceedingly sensitive. Indeed, due to their inherent feature of using different primers targeting different sequences (due to bisulphite treatment), MSPs are capable of amplifying and detecting even rare methylated alleles. Further improvements in sensitivity together with the possibility of quantifying methylated alleles have been achieved in quantitative PCR assays. One of these techniques, MethyLight, exploits the use of the combination of a methylation-specific primer and fluorescent-labelled TaqMan probes to detect and quantify methylated alleles [155]. In a modified version of MethyLight called QAMA (quantitative analysis of methylated alleles), the improved discriminative power of two different probes labelled with different fluorophores, namely, MGB modified Taqman probes, are used to measure methylated and unmethylated sequences in a single tube [156]. In another innovative real-time PCR variant, called HeavyMethyl, modified oligonucleotide blockers targeting unmethylated sites of primer annealing are used to increase the sensitivity and specificity of detecting and quantifying rare methylated alleles,. The binding of the blockers, preventing the amplification of the unmethylated targets, may be employed to detect very few copies of methylated sequence in serum samples with a very high sensitivity and specificity [157]. In MethylQuant assays, RT-qPCR is carried out without the use of expensive fluorescent labelled probes. Two distinct RT-qPCRs are simultaneously carried out, one using methylation-specific primers after bisulphate treatment and the other with wild-type primers without previous bisulphite treatment. In this setting, the difference in the cycle threshold of the two PCR reactions estimates the ratio between the methylated and the total target [158]. All the above-mentioned applications can assess and quantify the methylation state in 1 or a few genes/loci in each assay. Other applications exist which utilize PCR as a step for subsequent sequencing or microarray analysis intended for high throughput screening of thousands of targets simultaneously. Such applications require very expensive instrumentation for pyrosequencing or microarray preparation and analysis, and are therefore restricted for research use. Nevertheless, in this field, their use may allow the identification of candidate biomarkers whose full validation may be attained by the previously described PCR-based technique.

REFERENCES

[1]　Harris NL, Jaffe ES, Diebold J, Flandrin G, Muller-Hermelink HK, Vardiman J, Lister TA, Bloomfield CD. The World Health Organization classification of hematological malignancies report of the Clinical Advisory Committee Meeting, Airlie House, Virginia, November 1997. Mod Pathol 2000; 13: 193-207.

[2]　Rezuke WN, Abernathy EC, Tsongalis GJ. Molecular diagnosis of B- and T-cell lymphomas: fundamental principles and clinical applications. Clin Chem 1997; 43: 1814-23.

[3]　Hodges E, Williams AP, Harris S, Smith JL. T-cell receptor molecular diagnosis of T-cell lymphoma. Methods Mol Med 2005, 115: 197-215.

[4]　Lewin B. La diversità immunitaria. In Lewin B Ed. Il Gene VIII, Bologna, Zanichelli Editore:, 2006; pp. 782-817.

[5]　Tamura K, Yagihara H, Isotani M, Ono K, Washizu T, Bonkobara M. Development of the polymerase chain reaction assay based on the canine genome database for detection of monoclonality in B cell lymphoma. Vet Immunol Immunopathol 2006; 110: 163-.7

[6]　Yagihara H, Tamura K, Isotani M, Ono K, Washizu T, Bonkobara M. Genomic organization of the T-cell receptor gamma gene and PCR detection of its clonal rearrangement in canine T-cell lymphoma/leukemia. Vet Immunol Immunopathol 2007; 115: 375-82.

[7]　Jung D, Giallourakis C, Mostoslavsky R, Alt FW. Mechanism and control of V(D)J recombination at the immunoglobulin heavy chain locus. Annu Rev Immunol 2006; 24: 541-70.

[8]　Arun SS, Breuer W, Hermanns W. Immunohistochemical examination of light-chain expression (lambda/kappa ratio) in canine, feline, equine, bovine and porcine plasma cells. Zentralbl Veterinarmed A 1996; 43: 573-6.

[9]　Vernau W. Clonal rearrangements of antigen receptor genes in the diagnosis of lymphoid neoplasia. In ACVP and ASVCP Eds. Proceedings of 55th Annual Meeting of the American College of Veterinary Pathologists (ACVP) and 39th Annual Meeting of the American Society of Clinical Pathology (ASVCP), 2004; Middleton, WI, USA.

[10]　Vernau W, Moore PF. An immunophenotypic study of canine leukemias and preliminary assessment of clonality by polymerase chain reaction. Vet Immunol Immunopathol 1999; 69: 145-64.

[11]　Burnett RC, Vernau W, Modiano JF, Olver CS, Moore PF, Avery AC. Diagnosis of canine lymphoid neoplasia using clonal rearrangements of antigen receptor genes. Vet Pathol 2003; 40: 32-41.

[12]　Gentilini F, Calzolari C, Turba ME, Bettini G, Famigli Bergamini P. Genescanning analysis of Ig/TCR gene rearrangements to detect clonality in canine lymphoma. Vet Immunol Immunopathol 2009; 127: 47-56.

[13]　Wilkerson MJ, Dolce K, Koopman T, Shuman W, Chun R, Garrett L, Barber L, Avery A.Lineage differentiation of canine lymphoma/leukemias and aberrant expression of CD molecules. Vet Immunol Immunopathol 2005; 106: 179-96.

[14]　Lin J, Kennedy SH, Svarovsky T, Rogers J, Kemnitz JW, Xu A, Zondervan KT. High-quality genomic DNA extraction from formalin-fixed and paraffin-embedded samples deparaffinized using mineral oil. Anal Biochem 2009; 395: 265-7.

[15]　Lana SE, Jackson TL, Burnett RC, Morley PS, Avery AC. Utility of polymerase chain reaction for analysis of antigen receptor rearrangement in staging and predicting prognosis in dogs with lymphoma. J Vet Intern Med 2006; 20: 329-34.

[16]　Valli VE, Vernau W, de Lorimier LP, Graham PS, Moore PF. Canine indolent nodular lymphoma. Vet Pathol 2006; 43: 241-56.

[17]　Keller RL, Avery AC, Burnett RC, Walton JA, Olver CS. Detection of neoplastic lymphocytes in peripheral blood of dogs with lymphoma by polymerase chain reaction for antigen receptor gene rearrangement. Vet Clin Pathol 2004; 33: 145-9.

[18]　van Dongen JJ, Langerak AW, Brüggemann M, Evans PA, Hummel M, Lavender FL, Delabesse E, Davi F, Schuuring E, García-Sanz R, van Krieken JH, Droese J, González D, Bastard C, White HE, Spaargaren M, González M, Parreira A, Smith JL, Morgan GJ, Kneba M, Macintyre EA. Design and standardization of PCR primers and protocols for detection of clonal immunoglobulin and T-cell receptor gene recombinations in suspect lymphoproliferations: report of the BIOMED-2 Concerted Action BMH4-CT98-3936. Leukemia 2003; 17: 2257-317.

[19]　Avery PR, Avery AC. Molecular methods to distinguish reactive and neoplastic lymphocyte expansions and their importance in transitional neoplastic states. Vet Clin Pathol 2004; 33: 196-207.

[20]　Dictor M, Skogvall I, Warenholt J, Rambech E. Multiplex polymerase chain reaction on FTA cards vs. flow cytometry for B-lymphocyte clonality. Clin Chem Lab Med 2007; 45: 339-45.

[21]   Murphy KM, Berg KD, Geiger T, Hafez M, Flickinger KA, Cooper L, Pearson P, Eshleman JR. Capillary electrophoresis artifact due to eosin: implications for the interpretation of molecular diagnostic assays. J Mol Diagn 2005; 7: 143-8.

[22]   Werner JA, Woo JC, Vernau W, Graham PS, Grahn RA, Lyons LA , Moore PF. Characterization of feline immunoglobulin heavy chain variable region genes for the molecular diagnosis of B-cell neoplasia. Vet Pathol 2005; 42: 596-607.

[23]   Moore PF, Woo JC, Vernau W, Kosten S, Graham PS. Characterization of feline T cell receptor gamma (TCRG) variable region genes for the molecular diagnosis of feline intestinal T cell lymphoma. Vet Immunol Immunopathol 2005; 106: 167-78.

[24]   Henrich M, Hecht W, Weiss AT, Reinacher M.A new subgroup of immunoglobulin heavy chain variable region genes for the assessment of clonality in feline B-cell lymphomas. Vet Immunol Immunopathol 2009; 130: 59-69.

[25]   Weiss AT, Hecht W, Henrich M, Reinacher M. Characterization of C-, J- and V-region-genes of the feline T-cell receptor gamma. Vet Immunol Immunopathol 2008; 124: 63-74.

[26]   Breen M, Modiano JF. Evolutionarily conserved cytogenetic changes in hematological malignancies of dogs and humans-man and his best friend share more than companionship. Chromosome Res 2008; 16:145-54.

[27]   Jonsson OG, Kitchens RL, Scott FC, Smith RG. Detection of minimal residual disease in acute lymphoblastic leukemia using immunoglobulin hypervariable region specific oligonucleotide probes. Blood 1990; 76: 2072-9.

[28]   Calzolari C, Gentilini F, Agnoli C, Zannoni A, Peli A, Cinotti S, Famigli Bergamini P. PCR Assessment of Minimal Residual Disease in 8 Lymphoma-Affected Dogs. Vet Res Commun 2006; 30: (Suppl. 1) 285-8.

[29]   Yamazaki J, Baba K, Goto-Koshino Y, Setoguchi-Mukai A, Fujino Y, Ohno K, Tsujimoto H. Quantitative assessment of minimal residual disease (MRD) in canine lymphoma by using real-time polymerase chain reaction. Vet Immunol Immunopathol 2008; 126: 321-31.

[30]   Gentilini F, Turba ME, Calzolari C, Cinotti S, Forni M, Zannoni A. Real-time quantitative PCR using hairpin-shaped clone-specific primers for minimal residual disease assessment in an animal model of human non-Hodgkin lymphoma. Mol Cell Probes 2010; 24: 6-14.

[31]   Hazbón MH, Alland D. Hairpin primers for simplified single-nucleotide polymorphism analysis of Mycobacterium tuberculosis and other organisms. J Clin Microbiol 2004; 42: 1236-42.

[32]   Longley BJ, Tyrrell L, Lu SZ, Ma YS, Langley K, Ding TG, Duffy T, Jacobs P, Tang LH, Modlin I. Somatic c-KIT activating mutation in urticaria pigmentosa and aggressive mastocytosis: establishment of clonality in a human mast cell neoplasm. Nat Genet 1996; 12: 312-4.

[33]   London CA, Galli SJ, Yuuki T, Hu ZQ, Helfand SC, Geissler EN. Spontaneous canine mast cell tumors express tandem duplications in the proto-oncogene c-kit. Exp Hematol 1999; 27: 689-97.

[34]   Ma Y, Longley BJ, Wang X, Blount JL, Langley K, Caughey GH. Clustering of activating mutations in c-KIT's juxtamembrane coding region in canine mast cell neoplasms. J Invest Dermatol 1999; 112: 165-70.

[35]   Ma Y, Zeng S, Metcalfe DD, Akin C, Dimitrijevic S, Butterfield JH, McMahon G, Longley BJ The c-KIT mutation causing human mastocytosis is resistant to STI571 and other KIT kinase inhibitors; kinases with enzymatic site mutations show different inhibitor sensitivity profiles than wild-type kinases and those with regulatory-type mutations. Blood 2002; 99: 1741-4.

[36]   Webster JD, Kiupel M, Yuzbasiyan-Gurkan V. Evaluation of the kinase domain of c-KIT in canine cutaneous mast cell tumors. BMC Cancer 2006; 6: 85

[37]   Webster JD, Yuzbasiyan-Gurkan V, Kaneene JB, Miller R, Resau JH, Kiupel M. The role of c-KIT in tumorigenesis: evaluation in canine cutaneous mast cell tumors. Neoplasia 2006; 8: 104-11.

[38]   Hadzijusufovic E, Peter B, Rebuzzi L, Baumgartner C, Gleixner KV, Gruze A, Thaiwong T, Pickl WF, Yuzbasiyan-Gurkan V, Willmann M, Valent P. Growth-inhibitory effects of four tyrosine kinase inhibitors on neoplastic feline mast cells exhibiting a Kit exon 8 ITD mutation. Vet Immunol Immunopathol 2009; 132: 243-50.

[39]   Patnaik AK, Ehler WJ, MacEwen EG. Canine cutaneous mast cell tumor: morphologic grading and survival time in 83 dogs. Vet Pathol 1984; 21: 469-74.

[40]   Murphy S, Sparkes AH, Smith KC, Blunden AS, Brearley MJ. Relationships between the histological grade of cutaneous mast cell tumours in dogs, their survival and the efficacy of surgical resection. Vet Rec 2004; 154: 743-6.

[41]   Webster JD, Kiupel M, Kaneene JB, Miller R, Yuzbasiyan-Gurkan V. The use of KIT and tryptase expression patterns as prognostic tools for canine cutaneous mast cell tumors. Vet Pathol. 2004; 41:371-7. Erratum in: Vet Pathol 2004; 41:543.

[42]   Preziosi R, Morini M, Sarli G. Expression of the KIT protein (CD117) in primary cutaneous mast cell tumors of the dog. J Vet Diagn Invest 2004; 16: 554-61.

[43]   Pryer NK, Lee LB, Zadovaskaya R, Yu X, Sukbuntherng J, Cherrington JM, London CA. Proof of target for SU11654: inhibition of KIT phosphorylation in canine mast cell tumors. Clin Cancer Res 2003; 9: 5729-34.

[44]   Hahn KA, Ogilvie G, Rusk T, Devauchelle P, Leblanc A, Legendre A, Powers B, Leventhal PS, Kinet JP, Palmerini F, Dubreuil P, Moussy A, Hermine O. Masitinib is safe and effective for the treatment of canine mast cell tumors. J Vet Intern Med 2008; 22: 1301-9.

[45]   Letard S, Yang Y, Hanssens K, Palmérini F, Leventhal PS, Guéry S, Moussy A, Kinet JP, Hermine O, Dubreuil P. Gain-of-function mutations in the extracellular domain of KIT are common in canine mast cell tumors. Mol Cancer Res 2008; 6: 1137-45.

[46]   Isotani M, Ishida N, Tominaga M, Tamura K, Yagihara H, Ochi S, Kato R, Kobayashi T, Fujita M, Fujino Y, Setoguchi A, Ono K, Washizu T, Bonkobara M. Effect of tyrosine kinase inhibition by imatinib mesylate on mast cell tumors in dogs. J Vet Intern Med 2008; 22: 985-8.

[47]   London CA, Malpas PB, Wood-Follis SL, Boucher JF, Rusk AW, Rosenberg MP, Henry CJ, Mitchener KL, Klein MK, Hintermeister JG, Bergman PJ, Couto GC, Mauldin GN, Michels GM. Multi-center, placebo-controlled, double-blind, randomized study of oral toceranib phosphate (SU11654), a receptor tyrosine kinase inhibitor, for the treatment of dogs with recurrent (either local or distant) mast cell tumor following surgical excision. Clin Cancer Res 2009; 15: 3856-65.

[48]   Broudy VC. Stem cell factor and hematopoiesis. Blood 1997; 90:1345-64

[49]   Liao AT, McMahon M, London CA. Identification of a novel germline MET mutation in dogs. Anim Genet 2006; 37: 248-52.

[50]   Downing S, Chien MB, Kass PH, Moore PE, London CA. Prevalence and importance of internal tandem duplications in exons 11 and 12 of c-kit in mast cell tumors of dogs. Am J Vet Res 2002; 63: 1718-23.

[51]   Zemke D, Yamini B, Yuzbasiyan-Gurkan V. Mutations in the juxtamembrane domain of c-KIT are associated with higher grade mast cell tumors in dogs. Vet Pathol 2002; 39: 529-35.

[52]   Jones CL, Grahn RA, Chien MB, Lyons LA, London CA. Detection of c-kit mutations in canine mast cell tumors using fluorescent polyacrylamide gel electrophoresis. J Vet Diagn Invest 2004; 16: 95-100.

[53]   Riva F, Brizzola S, Stefanello D, Crema S, Turin L. A study of mutations in the c-kit gene of 32 dogs with mastocytoma. J Vet Diagn Invest 2005; 17: 385-8.

[54]   Webster JD, Yuzbasiyan-Gurkan V, Miller RA, Kaneene JB, Kiupel M. Cellular proliferation in canine cutaneous mast cell tumors: associations with c-KIT and its role in prognostication. Vet Pathol 2007; 44: 298-308.

[55]   Dank G, Chien MB, London CA. Activating mutations in the catalytic or juxtamembrane domain of c-kit in splenic mast cell tumors of cats. Am J Vet Res 2002; 63: 1129-33.

[56]   Isotani M, Tamura K, Yagihara H, Hikosaka M, Ono K, Washizu T, Bonkobara M. Identification of a c-kit exon 8 internal tandem duplication in a feline mast cell tumor case and its favorable response to the tyrosine kinase inhibitor imatinib mesylate. Vet Immunol Immunopathol 2006; 114: 168-72.

[57]   Webster JD, Yuzbasiyan-Gurkan V, Thamm DH, Hamilton E, Kiupel M. Evaluation of prognostic markers for canine mast cell tumors treated with vinblastine and prednisone. BMC Vet Res 2008; 4: 32.

[58]   Li J, Wang L, Mamon H, Kulke MH, Berbeco R, Makrigiorgos GM Replacing PCR with COLD-PCR enriches variant DNA sequences and redefines the sensitivity of genetic testing. Nat Med 2008; 14: 579-84.

[59]   de Lange T. Shelterin: the protein complex that shapes and safeguards human telomeres. Genes and Development 2005; 19, 2100-10.

[60]   Chen LY , Liu D , Songyang Z . Telomere maintenance through spatial control of telomeric proteins . Mol Cell Biol 2007; 27: 5898-909.

[61]   Allsopp RC, Chang E, Kashefi-Aazam M, Rogaev EI, Piatyszek MA, Shay JW, Harley CB. Telomere shortening is associated with cell division *in vitro* and *in vivo*. Exp Cell Res 1995; 220: 194-200.

[62]   Blasco MA. Telomeres and human disease: ageing, cancer and beyond. Nature Reviews Genetics 2005; 6: 611-22.

[63]   Kipling D, Cook HJ. Hypervariable ultra-long telomers in mice. Nature (lond.) 1990; 347: 400-2.

[64]   Nasir L, Devlin P, Mckevitt T, Rutteman G, Argyle DJ. Telomere lengths and telomerase activity in dog tissues: a potential model system to study human telomere and telomerase biology. Neoplasia 2001; 3: 351-9.

[65]   Yazawa M, Okuda M, Setoguchi A, Iwabuchi S, Nishimura R, Sasaki N, Masuda K, Ohno K, Tsujimoto H. Telomere length and telomerase activity in canine mammary gland tumors. Am J Vet Res 2001; 62: 1539-43.

[66]   McKevitt TP, Nasir L, Devlin P. Telomere lengths in dogs decrease with increasing donor age. J Nutr 2002; 132: 1604S-06S.

[67]   Cadile CD, Kitchell BE, Newman RG, Biller BJ, Hetler ER. Telomere length in normal and neoplastic canine tissues. Am J Vet Res 2007; 68: 1386-91.

[68]   McKevitt TP, Nasir L, Wallis CV, Argyle DJ. A cohort study of telomere and telomerase biology in cats. Am J Vet Res 2003; 64: 1496-9.

[69]   Argyle D, Ellsmore V, Gault EA, Munro AF, Nasir L. Equine telomeres and telomerase in cellular immortalisation and ageing. Mech Ageing Dev 2003; 124: 759-64.

[70]   Long S, Argyle DJ, Gault EA, Nasir L. Inhibition of telomerase in canine cancer cells following telomestatin treatment. Vet Comp Oncol 2007; 5: 99-107.

[71]   Royle NJ, Foxon J, Jeyapalan JN, Mendez-Bermudez A, Novo CL, Williams J, Cotton VE. Telomere length maintenance--an ALTernative mechanism. Cytogenet Genome Res 2008; 122: 281-91.

[72]   Kim NW, Piatyszek MA, Prowse KR, Harley CB, West MD, Ho PL, Coviello GM, Wright WE, Weinrich SL, Shay JW. Specific association of human telomerase activity with immortal cells and cancer. Science 1994; 266: 2011-5.

[73]   Cong YS, Wright WE , Shay JW . Human telomerase and its regulation. Microbiol Mol Biol Rev 2002; 66: 407-25

[74]   Nasir L, Gault E, Campbell S, Veeramalai M, Gilbert D, McFarlane R, Munro A, Argyle DJ. Isolation and expression of the reverse transcriptase component of the Canis familiaris telomerase ribonucleoprotein (dogTERT). Gene 2004; 336: 105-13.

[75]   Long S, Argyle DJ, Nixon C, Nicholson I, Botteron C, Olby N, Platt S, Smith K, Rutteman GR, Grinwis GC, Nasir L. Telomerase reverse transcriptase (TERT) expression and proliferation in canine brain tumours. Neuropathol Appl Neurobiol 2006; 32: 662-73.

[76]   Avilion AA, Piatyszek MA, Gupta J, Shay JW, Bacchetti S, Greider CW. Human telomerase RNA and telomerase activity in immortal cell lines and tumor tissues. Cancer Res 1996; 56: 645-50.

[77]   Yi X, Tesmer VM, Savre-Train I, Shay JW, Wright WE. Both transcriptional and posttranscriptional mechanisms regulate human telomerase template RNA levels. Mol Cell Biol 1999; 19: 3989-97.

[78]   Parkinson EK, Fitchett C, Cereser B. Dissecting the non-canonical functions of telomerase. Cytogenet Genome Res 2008; 122: 273-80.

[79]   Krupp G, Kühne K, Tamm S, Klapper W, Heidorn K, Rott A, Parwaresch R. Molecular basis of artifacts in the detection of telomerase activity and a modified primer for a more robust 'TRAP' assay. Nucleic Acids Res 1997; 25: 919-21.

[80]   Kim NW, Wu F.Advances in quantification and characterization of telomerase activity by the telomeric repeat amplification protocol (TRAP). Nucleic Acids Res 1997; 25: 2595-7.

[81]   Szatmari I, Tokes S, Dunn CB, Bardos TJ, Aradi J. Modified telomeric repeat amplification protocol: a quantitative radioactive assay for telomerase without using electrophoresis. Anal Biochem 2000; 282: 80-8.

[82]   Szatmari I, Aradi J. Telomeric repeat amplification, without shortening or lengthening of the telomerase products: a method to analyze the processivity of telomerase enzyme. Nucleic Acids Res 2001; 29: E3.

[83]   Hou M, Xu D, Björkholm M, Gruber A.Real-time quantitative telomeric repeat amplification protocol assay for the detection of telomerase activity. Clin Chem 2001; 47: 519-24.

[84]   Elmore LW, Forsythe HL, Ferreira-Gonzalez A, Garrett CT, Clark GM, Holt SE. Real-time quantitative analysis of telomerase activity in breast tumor specimens using a highly specific and sensitive fluorescent-based assay. Diagn Mol Pathol 2002; 11:177-85.

[85]   Wege H, Chui MS, Le HT, Tran JM, Zern MA. SYBR® Green I real-time telomeric repeat amplification protocol for the rapid quantification of telomerase activity. Nucleic Acids Res 2003; 31: E3-3.

[86]   Huang YP, Liu ZS, Tang H, Liu M, Li X. Real-time telomeric repeat amplification protocol using the duplex scorpion and two reverse primers system: the high sensitive and accurate method for quantification of telomerase activity. Clin Chim Acta 2006; 372: 112-9.

[87]   Yazawa M, Okuda M, Setoguchi A, Nishimura R, Sasaki N, Hasegawa A, Watari T, Tsujimoto H. Measurement of telomerase activity in dog tumors. J Vet Med Sci 1999; 61: 1125-9.

[88]   Funakoshi Y, Nakayama H, Uetsuka K, Nishimura R, Sasaki N, Doi K. Cellular proliferative and telomerase activity in canine mammary gland tumors. Vet Pathol 2000; 37: 177-83.

[89]   Cadile CD, Kitchell BE, Biller BJ, Hetler ER, Balkin RG.Telomerase activity as a marker for malignancy in feline tissues. Am J Vet Res 2001; 62: 1578-81.

[90]   Spangler EA, Rogers KS, Thomas JS, Pustejovsky D, Boyd SL, Shippen DE.Telomerase enzyme activity as a diagnostic tool to distinguish effusions of malignant and benign origin. J Vet Intern Med 2000; 14: 146-50.

[91]   Panarese S, Brunetti B, Sarli G. Evaluation of telomerase in canine mammary tissues by immunohistochemical analysis and a polymerase chain reaction-based enzyme-linked immunosorbent assay. J Vet Diagn Invest 2006; 18: 362-8.

[92]   Colitz CM, Barden CA, Lu P, Chandler HL. Expression and characterization of the catalytic subunit of telomerase in normal and cataractous canine lens epithelial cells. Mol Vis 2006; 12: 1067-76.

[93]   Carioto LM, Kruth SA, Betts DH, King WA. Telomerase activity in clinically normal dogs and dogs with malignant lymphoma. Am J Vet Res 2001; 62: 1442-6.

[94]   Hipple AK, Colitz CM, Mauldin GN, Mauldin GE, Cho DY. Telomerase activity and related properties of normal canine lymph node and canine lymphoma. Vet Comp Oncol 2003; 1: 140-51.

[95]   Renwick MG, Argyle DJ, Long S, Nixon C, Gault EA, Nasir L. Telomerase Activity and cTERT expression in canine lymphoma: correlation with Ki67 immunoreactivity. Vet Comp Oncol 2006; 4, 141-50.

[96]   Leri, A, Barlucchi, L., Limana, F., Deptala, A., Darzynkiewicz, Z., Hintze, T.H., Kajstura, J., Nadal-Ginard, B., Anversa, P. Telomerase expression and activity are coupled with myocyte proliferation and preservation of telomeric length in the failing heart. Proceedings of the National Academy of Science (USA) 2001; 98, 8626-31.

[97]   Kow K, Thamm DH, Terry J, Grunerud K, Bailey SM, Withrow SJ, Lana SE. Impact of telomerase status on canine osteosarcoma patients. J Vet Intern Med 2008; 22: 1366-1372.

[98]   Fusaro L, Panarese S, Brunetti B, Zambelli D, Benazzi C, Sarli G. Quantitative analysis of telomerase in feline mammary tissues. J Vet Diagn Invest 2009; 21: 369-73.

[99]   Zavlaris M, Angelopoulou K, Vlemmas I, Papaioannou N. Telomerase reverse transcriptase (TERT) expression in canine mammary tissues: a specific marker for malignancy? Anticancer Res 2009; 29: 319-25.

[100]   Mealey KL, Bentjen SA, Gay JM, Cantor GH. Ivermectin sensitivity in collies is associated with a deletion mutation of the mdr1 gene. Pharmacogenetics 2001; 11: 727-33.

[101]   Mealey KL, Fidel J, Gay JM, Impellizeri JA, Clifford CA, Bergman PJ. ABCB1-1Δ polymorphism can predict hematologic toxicity in dogs treated with vincristine. J Vet Intern Med 2008; 22: 996-1000.

[102]   Geyer J, Döring B, Godoy JR, Moritz A, Petzinger E. Development of a PCR-based diagnostic test detecting a nt230(del4) MDR1 mutation in dogs: verification in a moxidectin-sensitive Australian Shepherd. J Vet Pharmacol Ther 2005; 28: 95-9.

[103]   Geyer J, Klintzsch S, Meerkamp K, Wöhlke A, Distl O, Moritz A, Petzinger E. Detection of the nt230(del4) MDR1 mutation in White Swiss Shepherd dogs: case reports of doramectin toxicosis, breed predisposition, and microsatellite analysis. J Vet Pharmacol Ther 2007; 30: 482-5.

[104]   Mealey KL, Meurs KM. Breed distribution of the ABCB1-1Delta (multidrug sensitivity) polymorphism among dogs undergoing ABCB1 genotyping. J Am Vet Med Assoc 2008; 233: 921-4.

[105]   Lupu M, Sullivan EW, Westfall TE, Little MT, Weigler BJ, Moore PF, Stroup PA, Zellmer E, Kuhr C, Storb R. Use of multigeneration-family molecular dog leukocyte antigen typing to select a hematopoietic cell transplant donor for a dog with T-cell lymphoma. J Am Vet Med Assoc 2006; 228: 728-32.

[106]   Wagner JL, Burnett RC, DeRose SA, Francisco LV, Storb R, Ostrander EA. Histocompatibility testing of dog families with highly polymorphic microsatellite markers. Transplantation 1996; 62: 876–7.

[107]   Wagner JL, Works JD, Storb R. DLA-DRB1 and DLA-DQB1 histocompatibility typing by PCR-SSCP and sequencing. Tissue Antigens 1998; 52: 397–401.

[108]   Wagner JL. Molecular organization of the canine major histocompatibility complex. J Hered 2003; 94: 23-6.

[109]   Angles JM, Kennedy LJ, Pedersen NC. Frequency and distribution of alleles of canine MHC-II DLA-DQB1, DLA-DQA1 and DLA-DRB1 in 25 representative American Kennel Club breeds. Tissue Antigens 2005; 66: 173-84.

[110]   Wagner JL, Palti Y, DiDario D, Faraco J. Sequence of the canine major histocompatibility complex region containing non-classical class I genes. Tissue Antigens 2005; 65: 549-55.

[111]   Hardt C, Ferencik S, Tak R, Hoogerbrugge PM, Wagner V, Grosse-Wilde H. Sequence-based typing reveals a novel DLA-88 allele, DLA-88*04501, in a beagle family. Tissue Antigens 2006; 67: 163-5.

[112]   Kennedy LJ, Brown JJ, Barnes A, Ollier WE, Knyazev S. Major histocompatibility complex typing of dogs from Russia shows further dog leukocyte antigen diversity. Tissue Antigens 2008; 71: 151-6.

[113]   Graumann MB, DeRose SA, Ostrander EA, Storb R. Polymorphism analysis of four canine MHC class I genes. Tissue Antigens 1998; 51: 374-81.

[114]   Gluckman E, Rocha V. Donor selection for unrelated cord blood transplants. Curr Opin Immunol 2006; 18: 565-70.

[115]   Hwang WY, Ong SY. Allogeneic haematopoietic stem cell transplantation without a matched sibling donor: current options and future potential. Ann Acad Med Singapore 2009; 38: 340-6.

[116]   Graves SS, Hogan W, Kuhr CS, Diaconescu R, Harkey MA, Georges GE, Sale GE, Zellmer E, Baran S, Jochum C, Stone B, Storb R. Stable trichimerism after marrow grafting from 2 DLA-identical canine donors and nonmyeloablative conditioning. Blood 2007; 110: 418-23.

[117] Jochum C, Beste M, Zellmer E, Graves SS, Storb R. CD154 blockade and donor-specific transfusions in DLA-identical marrow transplantation in dogs conditioned with 1-Gy total body irradiation. Biol Blood Marrow Transplant 2007; 13: 164-71.

[118] Dickinson PJ, Roberts BN, Higgins RJ, Leutenegger CM, Bollen AW, Kass PH, Lecouteur RA. Expression of receptor tyrosine kinases VEGFR-1 (FLT-1), VEGFR-2 (KDR), EGFR-1, PDGFRa and c-Met in canine primary brain tumours. Vet Comp Oncol 2006; 4: 132-40.

[119] Polton GA, Brearley MJ, Green LM, Scase TJ. Expression of E-cadherin in canine anal sac gland carcinoma and its association with survival. Vet Comp Oncol 2007; 5: 232-8.

[120] Qiu C, Lin D, Wang J, Wang L. Expression and significance of PTEN in canine mammary gland tumours. Res Vet Sci 2008; 85: 383-8.

[121] Rankin WV, Henry CJ, Turnquist SE, Turk JR, Beissenherz ME, Tyler JW, Rucker EB, Knapp DW, Rodriguez CO, Green JA Identification of survivin, an inhibitor of apoptosis, in canine urinary bladder transitional cell carcinoma. Vet Comp Oncol 2008; 6: 141-50.

[122] De Maria R, Miretti S, Iussich S, Olivero M, Morello E, Bertotti A, Christensen JG, Biolatti B, Levine RA, Buracco P, Di Renzo MF. met oncogene activation qualifies spontaneous canine osteosarcoma as a suitable pre-clinical model of human osteosarcoma. J Pathol 2009; 218: 399-408.

[123] Klopfleisch R, Gruber AD. Increased expression of BRCA2 and RAD51 in lymph node metastases of canine mammary adenocarcinomas. Vet Pathol 2009; 46: 416-22.

[124] Mortarino M, Gelain ME, Gioia G, Ciusani E, Bazzocchi C, Comazzi S. ZAP-70 and Syk expression in canine lymphoid cells and preliminary results on leukaemia cases. Vet Immunol Immunopathol 2009; 128: 395-401.

[125] Abdullah-Sayani A, Bueno-de-Mesquita JM, van de Vijver MJ. Technology Insight: tuning into the genetic orchestra using microarrays--limitations of DNA microarrays in clinical practice. Nat Clin Pract Oncol 2006; 3: 501-16.

[126] Bhattacharya S, Mariani TJ. Array of hope: expression profiling identifies disease biomarkers and mechanism. Biochem Soc Trans 2009; 37: 855-62.

[127] Rao NA, van Wolferen ME, van den Ham R, van Leenen D, Groot Koerkamp MJ, Holstege FC, Mol JA. cDNA microarray profiles of canine mammary tumour cell lines reveal deregulated pathways pertaining to their phenotype. Anim Genet 2008; 39: 333-45.

[128] Rao NA, van Wolferen ME, Gracanin A, Bhatti SF, Krol M, Holstege FC, Mol JA. Gene expression profiles of progestin-induced canine mammary hyperplasia and spontaneous mammary tumors. J Physiol Pharmacol 2009; 60 Suppl 1: 73-84.

[129] Thomson SA, Kennerly E, Olby N, Mickelson JR, Hoffmann DE, Dickinson PJ, Gibson G, Breen M. Microarray analysis of differentially expressed genes of primary tumors in the canine central nervous system. Vet Pathol 2005; 42: 550-8.

[130] Nagy R, Sweet K, Eng C. Highly penetrant hereditary cancer syndromes. Oncogene 2004; 23: 6445-70.

[131] Garber JE, Offit K. Hereditary cancer predisposition syndromes. J Clin Oncol 2005; 23: 276-92.

[132] Lingaas F, Comstock KE, Kirkness EF, Sørensen A, Aarskaug T, Hitte C, Nickerson ML, Moe L, Schmidt LS, Thomas R, Breen M, Galibert F, Zbar B, Ostrander EA. A mutation in the canine BHD gene is associated with hereditary multifocal renal cystadenocarcinoma and nodular dermatofibrosis in the German Shepherd dog. Hum Mol Genet 2003; 12: 3043-53.

[133] Veldhoen N, Stewart J, Brown R, Milner J. Mutations of the p53 gene in canine lymphoma and evidence for germ line p53 mutations in the dog. Oncogene 1998; 16: 249-55.

[134] Veldhoen N, Watterson J, Brash M, Milner J. Identification of tumour-associated and germ line p53 mutations in canine mammary cancer. Br J Cancer 1999; 81: 409-15.

[135] Banerji N, Kapur V, Kanjilal S. Association of germ-line polymorphisms in the feline p53 gene with genetic predisposition to vaccine-associated feline sarcoma. J Hered 2007; 98: 421-7.

[136] Gentilini F, Turba ME, Forni M, Cinotti S. Complete sequencing of full-length canine ataxia telangiectasia mutated mRNA and characterization of its putative promoter. Vet Immunol Immunopathol 2009; 128: 437-40.

[137] Walther A, Johnstone E, Swanton C, Midgley R, Tomlinson I, Kerr D. Genetic prognostic and predictive markers in colorectal cancer. Nat Rev Cancer 2009; 9: 489-99.

[138] Smith J, Rattay T, McConkey C, Helliwell T, Mehanna H.Biomarkers in dysplasia of the oral cavity: a systematic review. Oral Oncol 2009; 45:647-53.

[139] Ogino S, Nosho K, Irahara N, Shima K, Baba Y, Kirkner GJ, Meyerhardt JA, Fuchs CS. Prognostic Significance and Molecular Associations of 18q Loss of Heterozygosity: A Cohort Study of Microsatellite Stable Colorectal Cancers. J Clin Oncol 2009; 27: 4591-8.

[140] Bønsdorff TB, Jansen JH, Thomassen RF, Lingaas F. Loss of heterozygosity at the FLCN locus in early renal cystic lesions in dogs with renal cystadenocarcinoma and nodular dermatofibrosis. Mamm Genome 2009; 20: 315-20.

[141] Yoshikawa Y, Morimatsu M, Ochiai K, Nagano M, Yamane Y, Tomizawa N, Sasaki N, Hashizume K. Analysis of genetic variations in the exon 27 region of the canine BRCA2 locus. J Vet Med Sci 2005; 67: 1013-7.

[142] Yoshikawa Y, Morimatsu M, Ochiai K, Nagano M, Tomioka Y, Sasaki N, Hashizume K, Iwanaga T. Novel variations and loss of heterozygosity of BRCA2 identified in a dog with mammary tumors. Am J Vet Res 2008; 69: 1323-8.

[143] Toda T, Oku H, Khaskhely NM, Moromizato H, Ono I, Murata T.Analysis of microsatellite instability and loss of heterozygosity in uterine endometrial adenocarcinoma. Cancer Genet Cytogenet 2001; 126: 120-7.

[144] Watanabe T, Wu TT, Catalano PJ, Ueki T, Satriano R, Haller DG, Benson AB 3rd, Hamilton SR.Molecular predictors of survival after adjuvant chemotherapy for colon cancer. N Engl J Med 2001; 344: 1196-206.

[145] McNiel EA, Griffin KL, Mellett AM, Madrill NJ, Mickelson JR.Microsatellite instability in canine mammary gland tumors. J Vet Intern Med 2007; 21: 1034-40.

[146] Garinis GA, Patrinos GP, Spanakis NE, Menounos PG.DNA hypermethylation: when tumour suppressor genes go silent. Hum Genet 2002; 111: 115-27.

[147] Plass C, Soloway PD. DNA methylation, imprinting and cancer. Eur J Hum Genet 2002; 10:6-16.

[148] Ting AH, McGarvey KM, Baylin SB.The cancer epigenome--components and functional correlates. Genes Dev 2006; 20:3215-31.

[149] Pelham JT, Irwin PJ, Kay PH. Genomic hypomethylation in neoplastic cells from dogs with malignant lymphoproliferative disorders. Res Vet Sci 2003; 74:101-4.

[150] Fosmire SP, Thomas R, Jubala CM, Wojcieszyn JW, Valli VE, Getzy DM, Smith TL, Gardner LA, Ritt MG, Bell JS, Freeman KP, Greenfield BE, Lana SE, Kisseberth WC, Helfand SC, Cutter GR, Breen M, Modiano JF. Inactivation of the p16 cyclin-dependent kinase inhibitor in high-grade canine non-Hodgkin's T-cell lymphoma. Vet Pathol 2007; 44: 467-78.

[151] Hiraoka H, Minami K, Kaneko N, Shimokawa Miyama T, Okamura Y, Mizuno T, Okuda M. Aberrations of the FHIT gene and Fhit protein in canine lymphoma cell lines. J Vet Med Sci 2009; 71: 769-77.

[152] Clark SJ, Harrison J, Paul CL, Frommer M.High sensitivity mapping of methylated cytosines. Nucleic Acids Res 1994; 22:2990-7.

[153] Brena RM, Huang TH, Plass C. Quantitative assessment of DNA methylation: Potential applications for disease diagnosis, classification, and prognosis in clinical settings. J Mol Med 2006; 84: 365-77.

[154] Herman JG, Graff JR, Myöhänen S, Nelkin BD, Baylin SB. Methylation-specific PCR: a novel PCR assay for methylation status of CpG islands. Proc Natl Acad Sci U S A 1996; 93: 9821-6.

[155] Eads CA, Danenberg KD, Kawakami K, Saltz LB, Blake C, Shibata D, Danenberg PV, Laird PW. MethyLight: a high-throughput assay to measure DNA methylation. Nucleic Acids Res 2000; 28:E32.

[156] Zeschnigk M, Böhringer S, Price EA, Onadim Z, Masshöfer L, Lohmann DR.A novel real-time PCR assay for quantitative analysis of methylated alleles (QAMA): analysis of the retinoblastoma locus. Nucleic Acids Res 2004; 32:e125.

[157] Cottrell SE, Distler J, Goodman NS, Mooney SH, Kluth A, Olek A, Schwope I, Tetzner R, Ziebarth H, Berlin K.A real-time PCR assay for DNA-methylation using methylation-specific blockers. Nucleic Acids Res 2004; 32:e10.

[158] Thomassin H, Kress C, Grange T. MethylQuant: a sensitive method for quantifying methylation of specific cytosines within the genome. Nucleic Acids Res 2004; 32:e168.

INDEX

A
Adenovirus, 80, 81, 82, 89
Allele-specific PCR, 59, 61
Amplicon sequencing, 18, 19
Annealing temperature, 18, 19, 22, 24, 41, 42, 44, 60, 121, 122
Antimicrobial resistance, 33, 34, 37, 38, 40-44, 46, 47
Antimicrobials, 33, 34, 39, 40, 47

B
Brucella, 59, 63-65, 71

C
Calicivirus, 81-83, 89
Cancer diagnosis, 106, 120
Clinical laboratory techniques, 3
Coronavirus, 80, 83-85, 89
Cryptosporidium parvum, 99, 100, 104

D
DNA sequencing, 33, 41, 44, 59, 62, 69, 71

F
Filariasis, 98, 102, 103
Flavivirus, 80, 84, 90
FRET, 21, 42, 45, 46, 62, 98, 100, 103

G
Genotyping, 13, 45, 59, 63, 98, 99, 101

H
Hematozoans, 98, 103
Herpesvirus, 80, 81, 83, 85, 90

I
Ig/TCR gene rearrangements, 106, 112
In-house assay, 18-20
Internal quality control, 18, 20, 26
Linearity, 18, 20, 27, 121, 122

L
Leptospira, 59, 64-66, 71
Lymphoma, 106, 109, 113, 126

M
Mast cell tumors, 106, 113
Mechanisms of resistance, 33, 37
Melting curve analysis, 18, 20, 25, 46, 62, 98, 101, 103
Minimal residual disease, 106, 111
Molecular diagnosis, 80, 90, 120
Multiplex PCR, 24, 42, 43, 59, 61, 65, 66, 68, 71, 72, 80, 81, 85, 99

www.ingramcontent.com/pod-product-compliance
Lightning Source LLC
Chambersburg PA
CBHW041714210326

41598CB00007B/651